GMO
Sapiens
The Life-Changing Science
of Designer Babies

GMO
Sapiens
The Life-Changing Science
of Designer Babies

Paul Knoepfler
University of California, Davis

World Scientific

NEW JERSEY · LONDON · SINGAPORE · BEIJING · SHANGHAI · HONG KONG · TAIPEI · CHENNAI · TOKYO

Published by

World Scientific Publishing Co. Pte. Ltd.

5 Toh Tuck Link, Singapore 596224

USA office: 27 Warren Street, Suite 401-402, Hackensack, NJ 07601

UK office: 57 Shelton Street, Covent Garden, London WC2H 9HE

Library of Congress Cataloging-in-Publication Data
Knoepfler, Paul, 1967–   , author.
   GMO sapiens : the life-changing science of designer babies / Paul Knoepfler.
     p. ; cm.
   Includes bibliographical references and index.
   ISBN 978-9814667005 (hardcover : alk. paper) -- ISBN 978-9814678537 (pbk. : alk. paper)
   I. Title.
   [DNLM: 1. Humans--genetics. 2. Organisms, Genetically Modified. 3. Cloning,
Organism--ethics. 4. Genetic Engineering--ethics. QH 442.6]
   QH442
   576.5--dc23
                              2015029803

British Library Cataloguing-in-Publication Data
A catalogue record for this book is available from the British Library.

Typeset by Stallion Press
Email: enquiries@stallionpress.com

Printed by FuIsland Offset Printing (S) Pte Ltd Singapore

This book is dedicated to some wonderful *Homo sapiens* including my family, friends, and students, and also to future *GMO sapiens* or human clones should they come to exist.

Preface

Writing *GMO Sapiens* was a wild ride.

People have been worrying about designer babies for decades. In some cases in the past, the anxiety reached a feverish level. In contrast, frankly most scientists did not take the past discussion of designer babies seriously enough to get concerned. I know that I didn't. It was, after all, just an idea and there were huge technical roadblocks in the way. A decade ago, many of us felt that it was more hype than anything else.

Today, scientists are taking the possibility of attempts to create designer babies very seriously. Those of us who have spoken out publicly on the issue generally oppose efforts at creating designer babies, but this is not true of everyone. Some prominent voices including certain scientists, ethicists, and legal scholars favor making designer babies. From all that I have seen lately, it is almost certain that someone will give it a try. Taking a stab at creation of designer babies today would be wildly unethical and dangerous, but since when has that kind of potential for disaster stopped people in the past from doing crazy stuff?

Scientists are paying close attention to this issue for another reason too. The same new genetic modification technology called CRISPR-Cas9 that has made attempts at creation of designer babies so readily possible has also made transformative new advances in laboratory science achievable too. Many scientists are very concerned that attempts at making designer babies by rogue laboratories could not only directly cause harm to individuals but also that such reckless efforts could derail other important and relatively non-controversial CRISPR-Cas9 gene-editing-based research in the process.

I believe that some scientists will try to make genetically modified (GM) people in the coming years. Science fiction is becoming reality in the biomedical world. This creates urgency for public education and more dialogue about how this technology could change the world and us with it. That change could be good, really bad, or some complicated combination of both. For these reasons, the idea for this book came to me with the goal being to make a positive difference in this area through furthering education and sparking discussion.

In writing this book, one major objective was to capture — in an understandable way — the nature of this tidal scientific and societal change for a broad audience and raise awareness of the possibility of GM people in the coming years. Education is a good thing, right? Still, some fellow scientists told me not to write this book as I began the process in late 2013 and early 2014.

"It will be too controversial and may unduly worry the public about biotechnology more generally," came the worries. "It will upset people."

Still, others were very supportive of the idea of a new book on human genetic modification. But I thought that it could not just be any book and definitely not a stodgy textbook. It had to be a book that was readable (and even enjoyable) for a large, diverse audience. A new, approachable book could be positive in terms of educating and inspiring a whole range of people on this topic, which was a secondary goal for this writing. I did not see any such book out there so I wrote this one. I felt the same way about the topic of stem cells when I started writing my first book, *Stem Cells: An Insider's Guide*. There was also a problematic gap in the stem cell world in 2012. Scientists reaching out and engaging with the public was needed on stem cells, and now the same is true about human genetic modification.

Encouragingly, some public discussion on the topic of human genetic modification has begun. For instance, I was happy to see as I continued writing this new book that in early 2015 some prominent genetic modification researchers such as Jennifer Doudna were publicly voicing concerns about human genetic modification. Doudna is one of the top scientists whose work has indirectly made human genetic modification possible through her research on CRISPR-Cas9 technology or "CRISPR" for short.

CRISPR makes relatively easy work of editing the genome just like one might do edits for a book using a computer keyboard. Much the same as those of us who write on keyboards often make mistakes, CRISPR can make mistakes too, but there is no autocorrect for a CRISPR-induced mistake in the human genome. Further, we may not even know the mistake happened until too late.

Doudna and other gene-editing scientists took the initiative to raise awareness of the possibility of CRISPR-mediated modification of human embryos. They started publicly calling for dialogue on human modification. The US Congress even held a hearing in June 2015 at which she was a panelist. Still more discussion bringing in as wide a public audience as possible would further help and a book could fulfill that role.

The more I learned as I researched this topic, the more I was inspired to blog about it and so I became a participant in the dialogue as well. For example, I reached out to leaders in the gene-editing and genomics fields for dialogue. I posted interviews with them on my blog, including Doudna as well as Harvard geneticist George Church and Duke legal scholar Nita Farahany. The latter two are both advocates from very different perspectives for future human modification under certain conditions. In May 2015 I was invited to speak at a Stanford Law School meeting on heritable human genetic modification organized by legal scholar Hank Greely.[i] I summarized that meeting here on my blog[ii] and it is discussed more in Chapter 10. We took questions from the public in the audience at the end of the meeting, which resulted in a great exchange of ideas.

Earlier I even found myself embroiled in a bit of an international incident with the UK equivalent of the US Surgeon General, Professor Sally Davies over human modification. Some in the UK, including Davies, were forcefully promoting approval of a form of heritable human

---

[i] http://stanford.io/1GcLZoT

[ii] http://www.ipscell.com/2015/05/stanfordhumangm/

genetic modification to try to prevent certain kinds of diseases called mitochondrial disorders. At that time, this issue was the subject of intense debate in the UK Parliament. She referred to the media reports of my concerns about safety risks of this so-called "three-person IVF" technology as "bunk." As much as that might sound negative, I took it as a very positive sign that there was a way to stimulate discussion, including via my blog, which could engage and include policy makers. Note that three-person IVF still has not been successfully achieved yet in the UK and is prohibited in the US by the FDA due to safety concerns.

Because I was in the mix on the topic of human genetic modification, my blog has become a resource on the topic. As a result, some sections of this book come in part or whole (e.g. interviews) from my blog.[iii]

The "GMO Sapiens" title of the book came from putting together "*Homo sapiens*" and "GMO" as a mashup. To be clear I would be opposed to calling actual GM people by the "GMO sapiens" nickname should they ever eventually be created, but we do need a way to refer to them right now even as they remain only hypothetical. This is important because even in their absence, they are important stakeholders in the discussion today.

The subtitle, "The life-changing science of designer babies" came about from discussions with my editors. For the purposes of this book a "designer baby" or "designer human" is one that has had a deliberate genetic modification introduced into their genome in a heritable manner for a specific purpose, where the goal of that design is medical or for enhancement of a trait.

Note that "life-changing" can mean either positive or negative changes, and a double meaning is intended there because genetic modification technology could literally change human lives for better or worse. This double meaning also reflects my efforts to maintain balance in discussing sometimes very divisive issues related to genetic modification.

Even though I went for a balanced discussion, I think my cautious nature becomes fairly clear as one reads this book. Since I love science and confess to being a bit of a techno nerd, I find genetic modification

---

[iii] http://www.ipscell.com

technology exciting. I admit that. My own team even uses this technology on human cells, but limited to the laboratory. However, I get deeply concerned when some talk about experimenting on human embryos that will become future children using a new, not entirely understood technology.

I did my best to avoid jargon as much as possible in writing. Where it proved essential to use terminology that not everyone may be familiar with, I have placed such words or phrases in a glossary in the back of the book. Please turn there if you find yourself puzzling over certain words.

There are also some partially or entirely new concepts here. In those cases, I have tried to carefully explain what I mean and include them in the glossary. For instance, the idea of "genetic tourism" is proposed in which people travel around the world to purchase genetic interventions that could even include the creation of GM children.

Another relatively new idea is "reproductive quarantine" in which governmental agencies forcibly prevent certain people from reproducing based on their genetics. This is similar conceptually to the forcible sterilization of past eugenics in the US and elsewhere including Nazi Germany, but more specifically reproductive quarantine as meant in this book applies primarily to GM people with modifications that have become perceived to be problematic.

Many people have helped make this book a reality so there are quite a few to thank. Darilyn Yap, Yugarani Thanabalasingam and Jane Alfred provided editing assistance. I want to thank my family, who put up with me working on it and provided helpful editing that, in some cases, included reading and editing the entire draft book. A big thanks is also due to Antonio Regalado and Nick Stockton who read the manuscript and provided insightful input. My UC Davis colleagues Mark Yarborough and Lisa Ikemoto were extremely helpful in providing feedback and information. I also want to thank Marcy Darnovsky and Jessica Cussins, who provided helpful perspectives and resources.

Every chapter has a number of pictures ranging from contemporary to historic photographs as well as illustrations and art even. Some of these images are photographs of people who are important players in the world of genetic modification. I included these specifically to put faces to the scientists and other people who played or continue to play key roles in making human genetic modification possible in one way or another, sometimes unintentionally. Some of the images are my own creations.

In the end I hope you enjoy this book and find it thought provoking. You may well find yourself feeling a range of emotions as you read. Hopefully, you will come away not only having learned new things, but also ready to discuss or even argue (maybe with me) on this topic.

As much as I am concerned about the possibility of GMO sapiens being created in the coming years, I do believe that if we speak out and take action we can greatly reduce that possibility. It is also important to encourage and advocate for useful laboratory research in the area of genetic modification of human cells and even in special cases of human embryos limited to test tube work.

In short, we can make a positive impact in the area of human genetic modification without throwing the baby out with the bath water.

Making a constructive difference is the whole point of this book.

Paul Knoepfler, Ph.D.
UC Davis School of Medicine
knoepfler@ucdavis.edu
@pknoepfler(Twitter)

Contents

Chapter 1

An Introduction to Playing God

"It was a fantastic achievement, but it was about more than infertility. It was also about issues like stem cells and the ethics of human conception. I wanted to find out exactly who was in charge, whether it was God himself or whether it was scientists in the laboratory ... It was us."

— Robert G. Edwards, July 24, 2003, on the occasion of the 25th birthday of the first "test tube" baby, Louise Brown, whom he helped to create.

Genetically modified (GM) human embryos

The world had a big shock this year. Researchers in China reported the creation of the first GM human embryos made using a new genetic technology, sparking a fiery controversy. Some have said that this kind of experimentation should stop, while others advocate going full steam ahead to make genetic modifications of people to try to prevent disease or even create enhanced, designer babies. I found myself in the middle of this as an outspoken scientist. With the conviction that the public needs to know much more about this unprecedented point in scientific and cultural history, I resolved to share it with you via this book.

You're only human ... but your kids could be more

Have you ever wished that there were something different about yourself? Maybe you imagined yourself taller, thinner, or stronger?

1

Smarter? More attractive? Healthier?

Or perhaps, as much as you love your children, you wished that there was something different about them. It is not that the love is missing, but it is precisely because you love them that you imagine they would be happier if they were different in some way.

It is also possible that some form of genetic disease runs in your family or a predisposition to cancer, Alzheimer's, or to some other potentially terrible health problem.

Until recently, you would have been able to do very little, if anything, about these situations, thoughts, and feelings. However, that might soon change. While you might not be able to fundamentally transform yourself or your existing children, in the near future, you just might be able to play God with your own new, little creations. Think of it as a very personal kind of experiment.

Technology that is already available today may well make this experimentation possible for anyone who can pay the price to make a new person, only one that is hoped to be "better." I mean a designer baby (Figure 1.1). You would be literally designing and producing a new type of baby via the same sort of technology that is used to make a GM tomato, mouse, or monkey.

The baby would be a genetically modified human or, to phrase it in an edgier manner, a GM human.

Would it be legal? In some places, yes.

Ethical? Hard to say, but I have my doubts.

Risky? Definitely.

Regardless of such thorny issues, it will be technically feasible to attempt, and you can bet that someone will try to do it in the coming years. The point of my bluntly laying out the incredible possibilities of what designer baby technology might be able to do for you was to illustrate how seductive it will be to many of us.

Should it fail at first, other scientists and doctors might be deterred. On the other hand, some could well see that as an opening to try it too. The technology will eventually become widely available. It might take two, five, or ten years, but it is coming. Should you as a parent do it? Many of us will answer, "yes."

Figure 1.1.   An artist's conception of a designer baby with some DNA bases, A's, C's, G's, and T's, represented by building blocks, changed by design in some ways. Image obtained from World Scientific Publishing Company (used with permission) and designed by the author. In part inspired by an image by Matt Collins.

Whether created for a medical reason or simply by parental choice, your new baby will be a novel human being with genetic modifications intended to improve him or her.

Yes, we are all special even if we were not born designer babies. And we are all unlike any human that has lived before us, thanks to the chromosomal reshuffling that occurs as a part of sexual reproduction. Even twins, although genetically identical, have unique differences in traits, or what we biologists call "phenotypes," based on, for example, the environment. However, your GM child will be unique not by chance, but rather by design. Further, this designer baby will be, at least in part, produced outside of the womb.

Your very own designer human will have a changed life compared to the existence that she or he would have experienced otherwise. Changed how? Your designer child, if all goes well, will be healthier or even simply "better" from your perspective as a parent. As I discuss later in this book, our perceptions of "better," may well reflect societal views on what constitutes desirable or superior traits. Your "better" child may not see herself or himself that way either. An important ethical question is whether a parent should be empowered to genetically alter a future child. This girl or boy would grow up to be literally a different person and one who had not consented to being genetically modified. Furthermore, all future members of that family could be GM people as well and again without consent. It is a tough question without a clear answer today.

Playing God via genetically changing human creation is made possible today by the marrying together of two powerful technologies. The first is now an old technique, in vitro fertilization (IVF), which was mastered by Nobel Laureate Robert Edwards and his colleague Patrick Steptoe four decades ago. The second is a new, cutting-edge genetic technology that makes it remarkably simple to directly tinker with the human genome (the DNA sequence) of an early embryo. When combined with IVF, these new genetic tools allow scientists to change the DNA, which is the blueprint of a human embryo, when it consists of just one or a few cells (for example, at the stage shown in Figure 1.2).

Edwards himself envisioned that IVF technology could have more powerful applications than for simply treating infertility. He also reasoned that with this power would come sweeping social implications [1]. You might say he was "all in" with the notion of playing God via IVF. He could read the handwriting on the test tube wall that scientists of the future would be able to make genetic modifications in people.

By the 1970s and 1980s, various kinds of non-human GMOs were in the works. The first ever GM animal, a mouse, was produced in 1974 by Professor Rudolph Jaenisch [2]. He inserted a virus' DNA sequence into a mouse's genome. Because this genetic change did not occur in the animal's sperm and eggs, Jaenisch's GM mouse could not pass this genetic change to its offspring. This meant it was not a heritable DNA change, but still this experiment was a huge scientific milestone, especially for the development of GM technology. In Chapter 2, you can read more about the birth of GM technology and the series of pioneering GM organisms that have been produced over

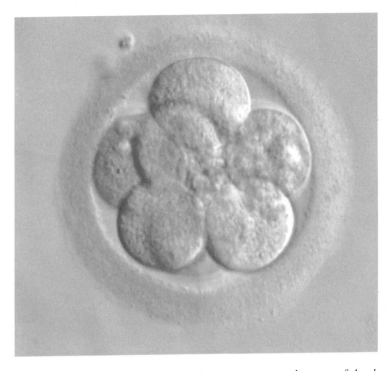

Figure 1.2. A normal human embryo is shown at a very early stage of development when it consists of just eight cells. Wikimedia, Rwjms IVF Program.

the years. There is a growing sense that GM humans are the next step in this progression.

Getting back to your new baby, he or she would be a person, but also, in a sense, a product. A specific company would make this unusual product and would charge you, and everyone else who elects to go this route, for the GM services it performs. There would be a profit generated and it likely would be a big one. It is not difficult to imagine a price tag of millions of dollars for a GM baby in the early days of this technology. If genetic modification of one's child-to-be catches on and it becomes trendy to have a designer baby, then the price would likely come down dramatically, while the possible impact to society — good or bad — would go up proportionately as more GM babies are produced.

Applying the terms GMO or GM to a new type of human being is no doubt controversial. However, they are accurate terms, despite the discomfort they might cause some.

If you go to the grocery store, you might pick up GM products without even realizing that you have done so. Personally I'm not worried about eating GMO containing foods per se, but I understand that there are some people who are very concerned about the cultivation and consumption of GM foods. Admittedly, I do try to get organic foods when possible for my family. Also, I don't use chemicals such as Roundup in my own home garden and haven't for many years. When you or I go to a plant nursery for garden supplies, could we unwittingly be picking up a GM plant?

In the future, you might find yourself going to a different type of nursery, a baby nursery, to pick up your GM baby. In this book, we will call this new hypothetical designer baby a "GMO sapiens" as a combination of GMO and *Homo sapiens*. For simplicity, we will most often leave the GMO sapiens name without italics.

Your GMO sapiens child might have avoided a terrible disease because genetic technology was used to correct a disease-causing mutation in a critical gene. Your baby, and you as its parent, may have literally dodged cystic fibrosis or a mutation in the *BRCA1* gene that puts women at elevated risk of breast and ovarian cancer, just to mention a couple of many possible examples. The hypothetical GM baby girl born without a *BRCA1* mutation would not only have a different life, but also she would never pass the mutation on to anyone in her future family tree.

This technology really is that powerful, and it is here, available right now. It is just that to be applied with hope of safety in humans, it would have to be tested for many more years and perfected. And today it has not yet been tested thoroughly and it is far from perfect. As a result, those contemplating using it in people would have to weigh the potential risks and benefits of using the genetic modification technology in the near future. One likely, very troubling consequence of trying to make GMO sapiens would be the byproduct of dozens or even hundreds of failed attempts in the form of diseased or deceased embryos, fetuses, and potentially even children. This is a truly disturbing thought. In addition, future experimental efforts to create GMO sapiens that get as far as implanting embryos into women also could lead to abortions caused by the experiments.

Another important limiting factor in human genetic modification is the open question of how to use this system to achieve a specific, desired positive outcome. While in some genetic diseases there is a clearly defined target such as the mutated cystic fibrosis gene that could be your focus for genetic modification, in many other cases there would not be a single clear bull's eye to go after. There could be a whole range of possible targets for genetic modification and to change a complex trait or disease, you would likely have to make many gene edits simultaneously. Therefore, to try to make a designer baby who is a genius, for example, would be a high-risk gamble today and for the foreseeable future. You might just as well end up making a kid who is less smart than she or he would have been without the genetic modification or who has autism or some other serious problem. In other words, the nature of the life-changing science applied to make your child could have made their existence far worse. We need to learn much more about the genetic basis of human traits before one could even hope to safely try to make changes in traits via genetics. Even then it may be ethically questionable.

Still, I expect some people will most likely go ahead and try anyway. By further discussion such as through this book, speaking out, and educating others, we can lower the risk that there will be disastrous, failed attempts to make GMO sapiens. For instance, it is crucial to point out that an already proven technology called preimplantation genetic diagnosis (PGD) works in most cases to prevent genetic disease. No editing is necessary because PGD works so well as a screening tool to select healthy embryos. You can read more on PGD in Chapter 5.

From one perspective, if many GM babies are born, it could be a giant step for humankind — perhaps as significant as the first step of a person on the moon. Would it be a step forward or back? Either way, this is a potentially life-changing and also a species-changing technology. It could catalyze a new phase of evolution for humanity if enough humans were born with genetic modifications (Figure 1.3). In fact,

Figure 1.3.   The evolution of humans, with *Homo sapiens* as the second to last on the right and *GMO sapiens* as yet another step in human evolution. Individuals depicted can be female or male. Adapted from free license image of José-Manuel Benitos on Wikimedia.

some of the people belonging to a movement called "transhumanism" are all for that new, "better" human reality made possible via genetic manipulations. They want humans of the near future to transcend from the relatively imperfect state of humans today to a new version of a person that is beyond human. Their symbol is h+, which is meant to symbolize that transcendence and some have named the better humans *Homo evolutis.*

Who does not want to be better? And what if you could make a baby who would grow up to be as smart and creative as Marie Curie or Albert Einstein? What if the next Stephen Hawking could be given the gift of never having had the catastrophic illness of amyotrophic lateral sclerosis (ALS)?

At the same time, there would be tremendous risks, both to individuals and to society, if we start to make designer babies. For each person made in this way, there would be a chance that errors are inadvertently introduced into other regions of their DNA, outside of the region targeted for the intended beneficial change. These glitches could sometimes lead to diseases, such as terrible developmental disorders, cancer, or death. Other kinds of errors could crop up as well, leading to similarly dire consequences.

There could even be insidious negative changes in GM humans that only manifest later in their life, such as changes in personality, leading, for example, to narcissism, violence, or suicide. Moreover, any new traits, whether good or bad, might forever be passed on to future generations because the introduced genetic changes would be inherited.

Some effects of making GMO sapiens could be totally unexpected as well and beyond our ability to predict. I discuss these dangers more in Chapter 6.

For society as a whole, the risks include a return to eugenics in a new incarnation, supercharged by novel genetic technologies that might have made eugenicists of the past jump up and down with excitement. We will return to transhumanism and eugenics in Chapter 7.

If you have what you perceive as a "perfect" GM baby, you could even potentially make a copy of her or him using cloning technologies, rather than rolling the dice on trying to make a different child (for a full discussion of human cloning, see Chapter 3).

The creation of people via technology rather than by sex may also fundamentally change the way we view our children, ourselves, and other people. Do we become an abstraction? A commodity? In other words, people could begin to be viewed a little bit more as things and somewhat less as human beings. The entire way we view parenthood could also change and lead to less of a parent-child bond.

In her groundbreaking book, *The Immortal Life of Henrietta Lacks* [3], author Rebecca Skloot tackles some of the deeply troubling issues related to use of people and their cells. Some of these same issues may well arise with human modification. Skloot also provides a powerful quote that resonates here from Nobel Laureate Elie Wiesel:

> "We must not see any person as an abstraction. Instead, we must see in every person a universe with its own secrets, with its own treasures, with its own sources of anguish, and with some measure of triumph."

Could genetic modification in some way reduce our humanity? In trying to use such technology to fight suffering from disease, for human enhancement, or in anti-aging efforts, might we change for the worse what it means to be human? Make ourselves more of a product?

Assuming one wants to give it a try despite the tremendous risks, how does one go about attempting to make a designer baby? For better or worse, it might not be that hard to try even if achieving success could prove to be elusive.

On the menu: IVF meets GMO

To produce a GMO sapiens baby, you would begin effectively by placing an order for her or him. It would be a team effort between you and the scientists involved. You might say it would "take a village and a lab" to make a GMO sapiens.

In the same way that today you might order a customized pizza with green olives, hold the onions, Italian ham, goat cheese, and a particular sauce, when you design and order your future GMO sapiens baby you could ask for very specific "toppings." In this case, toppings would mean your choice of unique traits, selected from a menu: green eyes, hold the diseases, Italian person's gene for lean muscle, fixed lactose intolerance so the designed individual can eat dairy, and a certain blood type.

Does this sound outlandish?

The personal genetics company, 23andMe, has already put together what is essentially a baby genetic predictor tool. For example, the company has specifically written about how one might go about, as a mother, selecting one's preference for green eyes and for a reduced risk of certain diseases, by screening the sperm of potential donors for these traits. For more about 23andMe and other companies developing tools to genetically design humans, see Chapter 5.

Another similar effort is underway from a company called GenePeeks,[i] co-founded by Professor Lee Silver of Princeton University, a proponent of human genetic modification. GenePeeks has developed a technology called Matchright. This service, available at some fertility clinics, enables customers to screen sperm from possible donors for how the genomes of those sperm when combined with the customer's might lead to certain

---

[i] https://www.genepeeks.com/

outcomes in possible future children. The search tool looks both for predicted disease risks and also specific traits.

At some point humans will not only want to use sperm or egg donor selection to make "better children," but also to turn to gene-editing technology that would be more direct and powerful. Gene edits are intentional changes made to the building blocks of life: the four molecules (so-called "bases") that make up DNA: adenine (A), cytosine (C), guanine (G), and thymine (T). For instance, a disease might be caused by a mutation that changes a G to a T in a particular gene. Gene-editing technology can be used to replace that T with a G, thus re-coding the DNA to now produce a normal version of that gene. Gene edits could also in principle be used to change traits that lead to human enhancement.

Other, bigger changes could include removing an entire disease-related gene (e.g. one with problems too complicated to address by small edits) and replacing it with a new, "healthy" copy from another source, such as a different person. Another change would be the insertion of a gene from another species. Adding in a gene from another type of organism would be a much more extreme change. The associated risks would be higher, but then again so might the intensity of the changed outcome.

Cutting edge technology: CRISPR-Cas9

These kinds of changes could be created in humans via a new gene-editing technology called CRISPR-Cas9 that has taken the life sciences by storm in just the last three years. I discuss CRISPR-Cas9 in more detail in Chapter 5, but a simple way to think of it now is like a genetic Swiss Army Knife that consists of the following: a magnifying glass (genome scanner), scissors (a special protein that cuts DNA) and a pencil (a cellular process that rewrites DNA bases) at the cut site, which gets glued back together.

CRISPR-Cas9 can pinpoint important but tiny gene sequences in our vast genomes, the genetic equivalent of finding a needle in a haystack. Once there, it can erase and/or change A's, C's, G's, or T's, or even larger genomic regions, in surprisingly precise ways. CRISPR can literally re-write the genomic book inside of us. However, it remains unknown how often it might go to the wrong page or paragraph, so to speak, or stay on the right page, but make an undesired edit there. CRISPR so far appears to be superior to an existing gene-editing technology called TALEN, but CRISPR is so new that the jury is still out on whether it will continue to be so strongly dominant in this area as it now appears.

With billions of bases in our genomes, even a very low error rate, yielding say an overall accuracy of 99.99… (and keep adding more 9s until you get tired) % could still have disastrous consequences, which is a source of great potential risk for GM babies. Many genetic diseases have relatively tiny mutations that cause severe disease or death. For instance, in Huntington's disease, there are just some extra copies of a repeating C-A-G DNA triplet. What this means is that a miniscule CRISPR mistake perhaps as small as a single DNA base pair unit, if it happens to fall in an important place in the genome, could cause severe disease or be fatal.

Where did CRISPR-Cas9 technology come from?

CRISPRs are genomic elements in bacteria that function as a source of immunity to future viral infection. CRISPR sequences are derived from viral DNA and are implanted into the bacterial genome as an immune memory. It is a bacteria's way of immunizing itself and its future progeny against viruses. Once tucked away, if in the future a bacterium is infected by the same virus that its ancestors had been exposed to, CRISPR allows it to more easily fight off the virus by literally chewing up the viral DNA. This digestion of the viral DNA is taken care of by another component of this system called Cas9 — the scissors of our genetic Swiss Army Knife. Cas9 is a special protein — an enzyme — that cuts DNA. Think of the CRISPR sequences as being like a magnifying glass or akin to a global positioning system (GPS) that guides the Cas9 to home in on the viral DNA for its destruction. Much the same way as the police have a database of the "fingerprints" of criminals, CRISPR elements act as a store of viral fingerprints that generations of bacteria keep and use to mount rapid immune responses to viral infections.

Following the discovery of this immune response system, researchers figured they could use CRISPR-Cas9 as a tool to edit the DNA of essentially any cell type, including that of humans. This could include the genomes of human reproductive cells and human embryos, leading to the creation of GM people that carry heritable, CRISPR-Cas9-introduced genetic changes.

After the design phase, the GM baby-to-be would go through a series of production steps, one or more of which might be completed outside of the womb, in a laboratory. IVF would play an important role. Since there are now currently scientists trying to produce artificial or laboratory-produced human wombs,[ii] it is even formally possible that, at some point in the future, the "production" of GMO sapiens babies could occur entirely outside of the human body.

Human reproduction could become a process nearly entirely independent of people, relying just on our cells. Scientists, once they had our cells, could "take it from there" so to speak. Not only would sex be unnecessary, but there could also be almost no physical parental involvement at all to produce a baby. The inclusion of a reproductive partner could eventually become optional too as technology advances further via cutting edge stem cell technology (more in Chapter 6).

Becoming a parent could turn into almost an intellectual exercise.

A project.

Instead of building a model airplane or jigsaw puzzle with your kid as a project, you as a parent would do a model building exercise, where your kid *is* the project. In place of plastic and glue or puzzle pieces, scientists would team up with you as the parent to make this new GM child, using your cells and genetic fabric as the starting material. The only other things needed from you would be the money to pay for the process and your input into the design of the baby.

For now, we will assume that your GM baby will be grown, for the most part in a real womb. Already for decades now, so-called "test-tube babies" have been produced without sexual intercourse and with one step — IVF — completed entirely outside the uterus in a laboratory. A test-tube baby produced by IVF has no alterations to its genes. At least we hope not, although there have been some concerns, as reported a few

---

ii http://bit.ly/1RvwlMg

years ago in *Scientific American,*[iii] of increased health risks and of genetic changes in individuals produced by IVF alone. However, IVF babies are generally healthy. Again as mentioned earlier in this chapter, with the power of the genetic screening of PGD, one can make the odds of IVF successfully creating a healthy baby even greater.

With GM baby production, the IVF step is simply a way to efficiently get sperm and eggs to do their thing together and to provide a window of time in which the laboratory scientists have access to the embryo to modify it. In the future production of GM babies, this IVF step would include changes to the DNA of the resulting human being. When that baby grows up and becomes an adult, they would then pass those same DNA changes to their own children as well, and so on. In this way, the collective genome of the entire human race (what we might call the "meta-genome") would be changed and there would likely be no going back.

How close are we to being able to produce GM babies today?

Designing and producing GM animals is fairly commonplace in science today. I predict that it will be for humans in the future as well, with potentially profound implications. Designing specific genetic changes in animals is something that I myself have done in the past as a researcher (in mice, in my case, see more about this in Chapter 6) and the same thing has been done countless times by other researchers too.

Why do we make GM animals? Like many other scientists, I create GM mice as tools to better understand specific processes in biology: the behavior of stem cells, the functions of particular genes, the way normal development happens, and how cancer happens too. The approaches we use to create GM mice are varied and sometimes complex so many of us are increasingly switching to the new, simpler CRISPR-Cas9 methodology. This technique focuses on editing genes in the reproductive cells (e.g. the eggs) of mice or in mouse embryos that still only consist of a single cell.

Traditionally, to make GM mice, researchers make changes to the genome of mouse embryonic stem cells. These cells, which also can be

---

[iii] http://bit.ly/1kSNWKL

made in human form in theory from any person, are like powerful shape shifters or "transformers" of the stem cell world. Embryonic stem cells can turn into any cell type in the body and hence can grow into a whole embryo. After genetically modifying the embryonic stem cells, these special cells are then transferred to female mice and develop into mouse embryos that grow into a GM mouse. In principle, this could be done in people too. However, with CRISPR-Cas9 technology it could be done even more simply without using embryonic stem cells.

The genetic modification step is most likely to be done in humans, either in the egg prior to fertilization or in the one-cell embryo right after its fertilization, using CRISPR-Cas9. It could even be done "earlier" in the human developmental spectrum in special kinds of stem cells that can turn into human sperm and eggs, called primordial germ cells or PGCs. By making gene edits via CRISPR-Cas9 in cells or embryos very early in development, all the cells of the resulting GM human body would probably carry the same desired gene edit. Otherwise, we could end up in a situation where the resulting GM baby's cells do not all have the same DNA sequence. This is called mosaicism, and it could lead to disease.

Your better baby

When the laboratory work and your final part (having the embryo implanted in your uterus, your partner's, or that of a surrogate) are all done, the end result would be your GMO sapiens baby. The hope would be that it would be a "better" baby than nature alone would have provided you with. What do I mean by better?

Clearly "better" is a subjective term and could invoke frightening scenarios, such as eugenics, the idea of producing genetically superior human beings and getting rid of genetically "inferior" ones. As I discuss in Chapter 7, in the past eugenics has led to disasters, such as the forced sterilization of thousands of people that occurred throughout the US. Powerful new gene-editing and reproductive technologies will not necessarily catalyze eugenics, but there is a risk that that could happen. It is a danger made greater by some people today embracing the idea of a new, benevolent eugenics that is empowered by novel technologies, such as CRISPR-Cas9.

I expect that, at first, the focus of heritable human genetic modification will be to design a healthier baby for you and ultimately a healthier adult. That sounds noble enough. For example, imagine a designer baby who is inherently resistant to a host of particularly nasty bacteria or parasites such as that which causes Malaria, or unable to be infected by certain viruses, such as a Hepatitis virus, Ebola, or HIV. A GMO sapiens made resistant to viral infection via CRISPR technology would be ironic given that bacteria use CRISPRs to resist viral infection too.

Or maybe the GM baby would have novel brain architecture or an innovative type of neuron, designed so that she or he could never get autism or Alzheimer's disease. Tinkering with genetics to change the architecture of the human brain, the most complex object known in the universe, would be fraught with danger. You might well end up causing cognitive impairments and brain diseases.

As mentioned above, the most likely first goal will be to create a GM baby that has been corrected for a single, disease-causing genetic mutation. This mutation, often normally passed along by you or your spouse to the child, would otherwise have caused her to be ill or to die. But now your baby would be born without this mutation, as it would have been corrected by gene-editing when she was just an embryo or even earlier in the reproductive cells used to make her.

Maybe at least in the early days of GMO sapiens production, those doctors, scientists and parents involved would avoid the temptation to tinker with "vanity" traits, such as height, musculature, skin or eye color, or even intelligence, pushing your child to score off the charts on an IQ test. They would just stick to making a healthier GM baby. Although if you paid enough money, perhaps you could make such, "a la carte," designer selections at certain businesses in some countries. It would cost more. This would be akin to the way you can pay extra for a "vanity" license plate in some countries (see for example, Figure 1.4).

I imagine that even the scientists involved in making GM humans would not appreciate people referring to these babies in that particular way, using the term "GM," despite probably taking great pride in the

Figure 1.4. GMO sapiens may be created in coming decades using adaptations of already available genomic, stem cell, cloning, and gene-editing technologies. Would the reason be for only health-related issues or also for more vanity-based choices, the way someone in Beverly Hills, California might order a vanity license plate?

precision of the genetic modifications they introduced into them. Some of the strongest advocates for making GM humans today already get defensive when the GM moniker is used to describe what they want to do, or should I say, what they want to create.

Are there people advocating for human genetic modification today? Yes, there are, although so far the advocates are in the minority amongst scientists. Still, pressure is growing around the world to let some forms of GM human production proceed. In a very basic form, it is probably doable as today's biotechnology is already at the point where scientists could try to genetically modify people. The advocates of human modification range from scientists to fertility doctors to ethicists. Dr. Tony Perry of the University of Bath in the UK went so far as to say that to *prevent* human embryo engineering would be unethical and a "sin of omission."[iv]

Those advocating for human genetic modification fall in two camps: those who support therapy to prevent serious genetic disease, and those futurists and some transhumanists who are keen to modify the human genome for the "betterment" of humanity. The latter are attracted to the new CRISPR-Cas9 technology and appear to support a new,

---

[iv] http://www.bbc.com/news/uk-politics-32633510

sometimes-called "liberal" eugenics, where the focus is on making people "better" rather than on preventing reproduction of "inferior" people. Within the former camp, some advocates have succeeded in getting the use of mitochondrial genetic modification technology (also known as "three-person IVF"; discussed more in Chapter 4) in humans legalized in the UK in 2015. Some advocates of three-person IVF argue that it would not lead to genetic modification, but science says otherwise. So proponents of some forms of human genetic modification are not just here today, but are also actively advocating its use now or in the future. Later in the book, I include my conversations with some of the leading skeptics and proponents of various forms of human genetic modification.

There are not just advocates of human genetic modification, but also a few dozen GM people walking around on planet Earth today. These GM people were created in the late 1990s to early 2000s via controversial fertility treatments that did not set out to make genetic modifications, but did so anyway. There is more on this fascinating story later in Chapter 4.

Getting back to you as a future parent-to-be, your GM embryo or fetus would ideally develop into an actual person, who could then lead an inherently healthier life. A better reality. Would existence really be better for that designer person? In turn, would society be better? Maybe. But then again, maybe not.

Since this is nearly entirely uncharted territory, how do we weigh the potential risks versus the kinds of mindboggling benefits that could come from producing GM people? For example, what would the risks be for the baby, the family, and for humanity? How would siblings treat a GMO sapiens brother or sister (Figure 1.5)? Could it change what it means to be a family?

These are not just hypothetical questions as today we are already speeding forward technologically to the point where people could be genetically modified in the next few years, although again the safety of doing such a thing is very much up in the air.

Advocates of making designer babies imagine new realities. For instance, Professor Silver of Princeton, despite advocating for designer babies, also imagines a future reality in which genetic modification technology changes society. In this future, Silver predicts an upper class

Figure 1.5. An imagined discussion between siblings in the future with GMO technology having been applied to people.

of the "GenRich," who are GM people that control society, and a lower class of "naturals" who are not.[v] In the predicted future of his book, *Remaking Eden*, Silver's GenRich become the new glitterati [4][vi]:

> "(In a few hundred years) the GenRich — who account for 10 percent of the American population — (will) all carry synthetic genes.... All aspects of the economy, the media, the entertainment industry, and the knowledge industry (will be) controlled by members of the GenRich class.... Naturals (will) work as low-paid service providers or as laborers.... (Eventually) the GenRich class and the Natural class will become... entirely separate species with no ability to cross-breed, and with as much romantic interest in each other as a current

[v] http://to.pbs.org/1djoeQb

[vi] http://www.geneticsandsociety.org/article.php?id=276

human would have for a chimpanzee.... (I)n a society that values individual freedom above all else, it is hard to find any legitimate basis for restricting the use of reprogenetics.... (T)he use of reprogenetic technologies is inevitable.... There is no doubt about it...whether we like it or not..."

In this future vision, it would appear a kind of Social Darwinism is at work turbocharged by genetics. Silver coined the term "Reprogenetics" to mean the coordinated use of assisted reproductive and genetics technologies to produce genetically enhanced humans. In a newer 2007 edition of his *Remaking Eden* book, Silver subtitled it, "How Genetic Engineering and Cloning will Transform the American Family."

GMO genesis

So how did we get here? Where did GM technology come from? What was the chain of events that brought us to the brink of producing GM people? It first started with simple, but groundbreaking, experiments where scientists started cutting and pasting pieces of DNA together, as well as mutating DNA in the laboratory using radiation.

This technology was then applied to living things, starting with simple organisms, such as bacteria and plants. GM mice followed. GMOs more generally, such as GM plants, have an intriguing and controversial history, which is the focus of Chapter 2.

The production of non-human GM organisms is currently accelerating at a rapid pace around the world. GMO production in laboratories and in the commercial sector has formed a foundation of knowledge and experience that is vital for science and medicine. They have also unintentionally established a technological foundation that makes the creation of new GMO sapiens now an eminently practical possibility.

It could happen in less than a decade, yet all the signs are that humanity is not ready for this decisive turning point in our history. One reason I have written this book is to educate others on this fascinating topic and spark discussion. There is still time to make a positive difference by preventing or delaying the creation of designer babies as well as by putting ethical and legal guidelines in place.

GMOs more broadly are one of the hottest, most controversial topics across the globe. GMOs today range from glow-in-the-dark green primates (Figure 1.6) [5] and other glowing organisms, to numerous forms of plants that have had their genetic makeup tweaked by scientists.

Many people hold strong opinions, both positive and negative, about GM plants and the related GM foods made from them, but surprisingly few appear to fathom how close we are to the point where actual GM people might be produced. This lack of awareness may explain why relatively few people are tuned in to this approaching, quite startling new reality. That will change soon. My sense is that it is better to spark discussion now rather than to wait until headlines trumpet either successful or failed attempts.

There are few legal or regulatory barriers to the production of GM people today in the US and the UK is going ahead with three-person IVF. In contrast, dozens of countries around the world prohibit any heritable human genetic modification. Although the production of GMO sapiens might be banned by one particular country, their creation could well still proceed in another with different laws. To my knowledge, for example, it

Figure 1.6. A GM marmoset named Wakaba. Its genome was modified to contain a green fluorescent protein (GFP) gene obtained from a jellyfish, which makes it glow in the dark when UV light is shone on it. This is shown in the lower left image, which shows Wakaba's green foot (on the right) next to that of a control. The lower right image shows a glowing green hair follicle. Image credit, Erika Sasaki; used with permission.

would not clearly be illegal at the federal level to create a GM baby in the US today and only certain American states have laws against it. However, the US House of Representatives is now considering a legal provision to restrict human embryo modification.

Even if it became illegal in the entire US, one could travel elsewhere, such as to China or perhaps to Central America. I foresee a new, troubling phenomenon of "genetic tourism" whereby people travel to other countries or states to purchase genetic modifications.

It might be helpful to address possible human genetic modification more broadly by talking about why researchers made GM plants and animals, as well as the risks involved in that arena. GM plant technology can offer particular advantages in a world undergoing climate change and famine. Such benefits would include plants with introduced drought resistance and the production of plants that have enhanced nutritional content. An example is the creation of genetically engineered "Golden rice" that produces nutritionally significant levels of vitamin A. However, there are concerns about the GMOs potentially having negative impact too and about the mega GMO corporations having so much control over what we can eat. Extending GM technology to humans could also lead to ethically treacherous attempts to make "Golden people," who — while not made of gold or even rich in vitamin A — would be intended to be far superior to the rest of humanity.

Another potential reason to produce GM people could be economic and that raises concerns as well, discussed further in Chapter 5. Fed by a desire for monetary gain, a GMO sapiens industry could emerge. Indeed, in a supply-and-demand economy, there could be sizable demand for GM human production. For instance, in a recent *NY Times* article on the scientist, Shoukhrat Mitalipov, who strongly advocates for human genetic modification via three-person IVF to prevent mitochondrial disorders, the desire of would-be families wanting to use his as-yet still unapproved assisted reproductive technology was described as having the "intensity of the hungry waiting for food."[vii]

[vii] http://nyti.ms/PMWOaF

So make no mistake, there will be people who will spend large sums of money to not only have a child, but also to make GM children that are better than those of their peers.

What happens to them and to society as a result?

If one looks at art and literature, collectively the prediction would be dire. There is a surprisingly long history of fictional works exploring humans hacking into their own creation. Almost without exception, the results imagined are dystopian in nature. Even today, polling suggests people are very concerned with the idea of human genetic modification and cloning. These wider issues are discussed in Chapter 8.

Some risk takers in the fertility industry today have shown interest in human cloning and genetic modification. While most likely the vast majority of fertility clinics would want nothing to do with human genetic modification, a few might view the creation of GMO sapiens as a potentially profitable side business to add to treating infertility. For example, they could say to an infertile couple, not only are we going to try to make you a baby after all these years, but also for an additional fee, we can make your baby "better" than it would have been otherwise.

A significant number of people would not be able to resist that temptation. Amongst those of us who are parents, who does not like to think that their kids are better than average? What if you could almost guarantee that your child would outshine all of the others in the neighborhood? All the kids in the country? Many of us would give in to that temptation.

The stem cell and cloning connections

The increasing potential to create GM people is being catalyzed not only by efforts to make GM plants and animals, but also indirectly and mostly unintentionally by genetics, genomics, reproductive, and stem cell research. For example, in 2013 for the first time ever, a team of researchers in Oregon led by Mitalipov cloned human embryonic stem cell (ESC) lines [6]. What this means is that his laboratory made human ESCs that are genetically identical to a human without the involvement of fertilization or any other natural reproductive process. Notably, several other research teams have now been able to reproduce this so-called human "therapeutic cloning" of

ESCs, including from a cell from an aged donor.[viii] Mitalipov is one of the key scientists who could play a role, intentional or not, in future human genetic modification.

The goal of this type of therapeutic cloning done by Mitalipov is to create customized, genetically matched human ESC lines that might be used to treat patients in the future with personalized stem cell-based medicines. These would be your very own genetically identical cells, but in an embryonic healthy form even if you might currently be sick or old. Although "therapeutic cloning" is the name that has stuck for the process of making ESCs via cloning, to date there is no clear prospect of using this technology for an actual therapy. Still, great medical benefits could come from this technology in the future.

Unfortunately, with just a few adjustments, others could use the same therapeutic cloning technology as part of the process of "reproductive" cloning, in which the goal is to make actual babies who have a genetic makeup that is identical to an existing or deceased person. Think cloning Elvis or the cloning portrayed in Star Wars.

As such, 2013 was a watershed year for human cloning, and potentially for human genetic modification moving much closer to becoming a reality. Even the therapeutic human cloning achieved in 2013 makes the production of GMO sapiens far more achievable, even though that is not the goal of the researchers who did the work; actually, in some cases, they strongly oppose reproductive cloning.[ix] Despite many scientists' opposition to reproductive cloning, other scientists co-opting that technology might misuse it, and during the human cloning process, it could be relatively easy to try to tweak a future baby's genetic makeup using CRISPR-Cas9.

Gene therapy is another area of biomedical science that has the potential to be used, in some people's view, to create GMO sapiens. The goal of gene therapy is to correct genetic defects that cause disease and in so doing reduce human suffering. In principle, the idea is that scientists and doctors reverse these mutations using genetic tools to replace the genetic mutation with normal DNA instead. Gene therapy is distinct from

---

[viii] http://bit.ly/1DZwtqm
[ix] http://bit.ly/1GWp4u9

making designers babies because it is done in such a way that it cannot be inherited by future generations. For that reason, I do not consider gene therapy patients to be GMO sapiens.

In this book, more broadly I discuss the science that is making the attempted creation of GM people a likely possibility in coming years, the pros and cons of GMO sapiens experiments, and the new future ahead of us. We need to educate ourselvers and others as well as have dialogue about the notion of making GMO sapiens ideally years before anyone would try to make them. Time is short though. The pressing need for discussion is a main reason why I wrote this book.

In the next chapter, I discuss the birth and growth of GMO technology. The next step could be GM people.

Are we ready?

References

1. Edwards, RGDJ Sharpe (1971) Social values and research in human embryology. *Nature*. 231:5298:87–91.
2. Jaenisch, RB Mintz (1974) Simian virus 40 DNA sequences in DNA of healthy adult mice derived from preimplantation blastocysts injected with viral DNA. *Proc Natl Acad Sci U S A*. 71:4:1250–4.
3. Skloot, R (2010) *The Immortal Life of Henrietta Lacks*. Crown Publishers.
4. Silver, LM (1997) *Remaking Eden: Cloning and Beyond in a Brave New World*. 1st ed. Avon Books.
5. Sasaki, E, *et al*. (2009) Generation of transgenic non-human primates with germline transmission. *Nature*. 459:7246:523–7.
6. Tachibana, M, *et al*. (2013) Human embryonic stem cells derived by somatic cell nuclear transfer. *Cell*. 153:6:1228–38.

Chapter 2

The Birth and Explosive Growth of GMOs

"You don't have to lick your finger to tell which way the wind is blowing. Like a spring tornado, this season's enthusiasm for genetic engineering is sweeping over agriculture, one of America's biggest businesses..."

— Ann Crittenden, New York Times, 1981.[i]

GM plants sprout

Scientists in a garage in Davis, California, had an audacious idea in 1981. It was one of those rare light bulb moments with the potential to change the world. This high-tech "eureka!" event happened not too far from where I am writing the book you are holding right now. I live in Davis, and am a professor at UC Davis School of Medicine.[ii]

Was it a vision for a new type of computer, like the idea for an Apple Mac from Steve Jobs, or a notion of some type of new black hole from Stephen Hawking? No, but this new technology would ultimately create something equally historic with major impact for science and society.

A new, commercial life form.

---

[i] http://nyti.ms/1AOb43y
[ii] http://www.ucdmc.ucdavis.edu/cellbio/faculty/knoepfler/

These innovators wanted to create a designer plant, one intended to be superior to those occurring naturally. The new plant would be produced by genetic engineering in a laboratory, with the intention that it be bought and consumed by the public. Plant scientists including those in that garage, also right here at UC Davis, and in other places around the world had begun brainstorming and experimenting with the idea of making the first commercially viable GM plants for human consumption. Farmers were also interested in making GM livestock as well, and that too has become a big business (as parodied in the cartoon by Liam Walsh in Figure 2.1).

The thinking at the time was that these new plants could both advance science and generate a profit. This innovation worked and catalyzed arguably the most significant change in agriculture in thousands of years.

Figure 2.1. Cartoon originally printed in the May 4, 2015 edition of the *New Yorker* by Liam Walsh. Reprinted here with permission.

Conventional plant breeding is a slow process. It entails waiting perhaps many years to breed traits into or out of plants, using a somewhat random process of trial and error — the same one that Gregor Mendel might have used almost two centuries ago to breed his famous peas. In the late 20th century, however, agricultural scientists had genetic engineering techniques at their disposal that could make "designer plants" in just a few years.

Out of that garage-based effort, a pioneering agricultural biotech startup came into being: Calgene. The name was chosen as a mashup of California and gene. UC Davis Professor of Agronomy and Range Science, Ray Valentine, teamed up with venture capitalist Norm Goldfarb to found Calgene They were a dynamic duo trying to turn the innovative mixture of plants, genetics, and marketing into agro-gold:

> "Goldfarb and Valentine began meeting on Saturday mornings, with Valentine tutoring Goldfarb in biology and genetics and Goldfarb talking about economics and financing. Within a few weeks, Goldfarb had quit his job at Intel and was ready to start his new company, Calgene. Valentine became vice president for research and development and formed a science advisory board from fellow scientists and other faculty members."[iii]

The focus of Calgene was to genetically engineer plants to improve them and to advance science. As with any biotech startup, there was also the goal of making large amounts of money by commercializing their products. It was not long before Calgene turned its sights on this second goal, making a profitable GMO food product.

Where did GMOs come from?

Where was the source of the inspiration to genetically modify plants, and in particular food crops?

Rewind a half dozen years from 1981, back to 1975 to the pre-Calgene era, and we find Valentine attending a historic meeting on the

---

[iii] https://localwiki.org/davis/Calgene_LLC

use of recombinant DNA technology at the Asilomar Conference Center in California. This "Asilomar meeting" as it became known would alter the course of DNA science, particularly when it was used to change living things with the purpose of making GMOs. Recombinant DNA means new hybrid forms of DNA made by scientists, sometimes in a test tube and other times in actual organisms. In the early years of this type of research in the 1970s, there was particular concern that GMOs made via recombinant technology such as pathogens could have profound negative consequences for the world. At the same time scientists were enthralled by the potential power for good from recombinant technology.

The point of the Asilomar DNA meeting was for leaders in the field to have an informed discussion about how to best handle the exploding potential and excitement of GMOs, as well as the concerns surrounding the dangers they might bring. The meeting was catalyzed by the realization among scientists and policymakers of the power of recombinant DNA technology, such that an immediate discussion was needed to determine how to both minimize risks and facilitate potentially powerful and beneficial research. There was a sense of urgency because the creation of GMOs had already begun. In a whitepaper (a written policy statement often drafted after a meeting), the meeting attendees called for proceeding with this research, but with caution [1, 2]:

> "The new techniques, which permit combination of genetic information from very different organisms, place us in an area of biology with many unknowns. Even in the present, more limited conduct of research in this field, the evaluation of potential biohazards has proved to be extremely difficult…the participants at the Conference agreed that most of the work on construction of recombinant DNA molecules should proceed provided that appropriate safeguards, principally biological and physical barriers adequate to contain the newly created organisms, are employed."

In other words, "let's go ahead, but keep GMOs locked up for now."

Valentine reportedly likened himself to just a "farmer"[iv] attending the Asilomar meeting with a slew of Nobel Laureate scientists, but it was clear

---

[iv] http://dateline.ucdavis.edu/dl_detail.lasso?id=7755

that he had something special to bring to the table in the form of a vision for plant biotechnology based on genetics. At the meeting, he wrote on the board "Molecular Farming" and the bigwigs loved the idea.

His experience at Asilomar may have been what prompted him a few years later to form Calgene to try to make GMO plants. The hope was to make money and, ultimately as the technology expanded, to revolutionize agriculture. It was an inspired idea, and the technology they used turned out to work even if paradoxically Calgene itself no longer exists due in large part to economic rather than GMO-based reasons.

More broadly, genetic modification of plants and animals has its origin in the pioneering genetic research of the 1970s and 1980s. During this revolutionary period, biomedical researchers were learning how to manipulate DNA in the laboratory and then, in turn, to make these genetic changes in living cells and in whole organisms. Tinkering with a code like DNA is really the only effective way to understand it well and to reveal the implications of that code for biological functions. Maybe, tinkering can also teach you on how to make new, useful forms of DNA.

In these early days of genetics and molecular biology, researchers increasingly began to produce cells and organisms with genetic modifications. This turned out to be a remarkably powerful way for scientists to make discoveries about gene functions, as well as to delve into cell and developmental biology and human disease in ways that were simply not possible in the past. This new approach of hacking into the DNA code and rewriting it was revolutionary.

How did this early research play out and what can it teach us about potential human genetic modification?

Studying the biological effects of changing the DNA code of model organisms, such as that of bacteria, yeast, flies, and mice, became a standard research approach decades ago. The key to making GM mice for studies of gene function and developmental biology was the discovery of mouse embryonic stem cells (ESC; recall these powerful cells that can make any cell in the human body) in the 1980s by Martin Evans, Matthew Kaufman, and Gail Martin, the last of which coined the term "embryonic stem cells."

Scientists could introduce mutations into specific mouse genes, using ESCs in the laboratory, and then grow entire mice from these modified ESC by injecting them into mouse embryos at a very early stage of their development. These modified embryos were then implanted into female mice. Once born, further screening was used to identify the mutants from the normal offspring. Scientists then had a new mutant mouse strain. A GMO. In principle, scientists could do much the same thing with human ESCs to create a GM human in the future.

Plant researchers started making mutant forms of plants too, but in part since there are no ESCs in plants different approaches were used. However, researchers studying bacteria were ahead of the GMO game, and the first ever GMO patent application of any kind was for a GM microbe filed in 1971.

Recall that the Asilomar DNA Conference calling for caution and a prudent path in future genetic engineering took place four years later in 1975. Genetic engineering was already happening years earlier. Perhaps the Asilomar meeting was a bit late, but it still achieved important outcomes such as responsible use of the technology.

The 1971 patent application for this GM microbe was for a bacterium that had an increased appetite for gobbling up oil. A type of bacteria that could digest oil was a brilliant idea given the increasing problem of oil spills. However, could the GM bacteria have some unintended consequences once released into the environment? Nobody knew.

The work of scientist Ananda Chakrabarty,[v] who had produced the new GM bacterial species — a version of Pseudomonas putida[vi] named "multi-plasmid hydrocarbon-degrading Pseudomonas" — was the basis for the patent. The irradiation and importantly the addition of circular bits of DNA called "plasmids" caused heritable mutations to occur in the DNA inside the bacteria and as a result, changed its biological properties. One such change was its affinity for eating oil.

---

[v] http://en.wikipedia.org/wiki/Ananda_Mohan_Chakrabarty
[vi] http://en.wikipedia.org/wiki/Pseudomonas_putida

It turns out that radiation alone had been used for quite some time to genetically modify organisms beyond what Chakrabarty did. Radiation causes random mutations by damaging DNA. During repair or copying of damaged DNA, mutations are introduced. In the 1960s, as part of the great enthusiasm for research involving radioactivity, scientists at the International Atomic Energy Agency and elsewhere routinely bombarded plants with radiation to see what would happen. The mutant plants that came out of these studies were not radioactive themselves, but on rare occasions they possessed new properties and sometimes those novel traits were quite useful or desirable. We might think of this as the "atomic gambler" approach to plant genetic modification. Very random. Still, it has had some notable successes, including useful variants of rice produced via radiation mutation studies right here in Davis, California, in 1971, and certain forms of red grapefruit,[vii] just to cite two examples. The same kinds of studies were done with bacteria to induce new traits.

Eventually, the Chakrabarty patent for the new and partially radiation-produced GM bacteria was granted. Perhaps not surprisingly, the Chakrabarty patent was challenged and this challenge led to a landmark US Supreme Court Case, Diamond v. Chakrabarty. The key question at issue was whether one could patent a living thing, such as a GM bacterial strain.

The Supreme Court voted 5–4 in favor of Chakrabarty, and the patent ended up becoming active practically speaking in 1980.[viii] At that time, the notion of genetic modification of any organism was so new that it sparked both excitement and fear. There was a sense in some quarters that GMOs could be like Frankenstein monsters.

Shortly after the ruling, on June 18, 1980 the *Washington Post* ran a cartoon by Herb Block (see Figure 2.2) that was meant to convey the sense of concern that this case could open the door to human genetic modification and the creation of Frankenstein-type monsters.

[vii] http://nyti.ms/1QCW1mm

[viii] https://supreme.justia.com/cases/federal/us/447/303/case.html

Figure 2.2. Editorial Cartoon by Herblock on the US Patent Office granting the patent for the first GMO, the GM bacterium that eats oil. Reprinted with permission from the Herb Block Foundation.

American historian of science Daniel Kevles articulated the concerns in this way [3]:

"However narrow the Court's decision, it seemed in many quarters to link the making of money to the making of monsters or, at least, to manipulation of the essence of life. As such, in the arena of biotechnology, it invited moral argument into the dynamic of patent deliberation for the first time in the history of American patent law and with consequences that remain to be seen."

The take home message from the court decision is that you can patent life as long as it is not naturally occurring life. In the long run this case, allowing GMOs to be considered intellectual property that can be owned, opened the door to the patenting of almost any GMO, and it paved the way for for-profit companies to see GM technology as a powerful potential new income stream.

This momentum for GMOs was further bolstered by the US FDA's approval two years later in 1982 of the first human GMO product: insulin made from GM bacteria that had been designed in a laboratory to produce large amounts of the drug.

Could one patent a GMO sapiens? No, due to various laws. What about patenting a *technique* for making GMO sapiens? This technology is likely to be patentable, and indeed patents for technologies that could in the future be used in the process of making GM humans are pending (e.g. Mitalipov's three-person IVF technique) and some have even been awarded including Professor Feng Zhang's CRISPR-Cas9 patent. Zhang is currently in a patent dispute with Professor Jennifer Doudna, who also has done pioneering CRISPR work, over the intellectual property rights to CRISPR.

The race for GM crops

Getting back to Calgene, could they succeed in creating a commercially viable GMO? Would some other researchers beat them to the punch? The race was on!

Between the time that Calgene was born and then later bought out by agro-giant Monsanto (Figure 2.4), the company developed a pipeline of several GMOs. Three types of plants were chosen to be the focus of these new GMO efforts: cotton, canola oil, and tomatoes. Again, keep in mind that in 1981 there were no GMO whole food plants — such as GM tomatoes — on the market or being eaten by people.

The cultivation of tomatoes was a particularly inspired area to focus on economically and culturally. Tomatoes represent a multi-billion dollar crop just in the US alone. Here, in Davis and Sacramento, the latter city nicknamed "The Big Tomato," the tomato business is of particular importance. Where I live in Davis, every summer I see streams of farm trucks roll down the dusty country roads and small highways carrying tons of tomatoes.

Strikingly, in 1986 only a few years after the GM bacteria patent, Washington University Professor Roger Beachy (now head of the World Food Center here at UC Davis since 2013) published a pioneering article on creation of a GMO tomato resistant to a harmful virus.[ix] Around the same time, Calgene[x] asked the US Department of Agriculture to approve a field test of a genetically engineered form of tobacco [4]. The tests were approved.

It was just six years later in 1994 when Calgene achieved another GMO milestone. People started to eat the first GMO whole food, Calgene's Flavr Savr tomato. Calgene had engineered Flavr Savr to resist spoiling, and applied for FDA approval to sell it to the public in 1992.

How did this GM anti-spoiling technology work?

Research had implicated a specific gene called ACC in fruit ripening, including in tomatoes. California plant researcher, Athanasios Theologis did some of the key research (Figure 2.3).

The scientists figured out that if one could inhibit ACC then perhaps fruits would have delayed ripening and, importantly, be slower to spoil such as by rotting. The approach used to inhibit ACC was called antisense technology, which involves flipping a gene around such that it produces an opposite, antagonizing form of itself that is inhibitory. The antisense against ACC worked. Theologis and colleagues discussed the success in a 1992 review article [5].

Imagine a tomato or other kinds of fruit that would not spoil for a long time, perhaps a few weeks at room temperature. It could become a commercial bonanza. However, there was a sense of caution at the time that GM plants should be monitored and, indeed, one year later a research article was published, which reported an assay for detecting Flavr Savr plants potentially growing where they should not be [6].

[ix] http://www.ncbi.nlm.nih.gov/pubmed/3457472

[x] https://localwiki.org/davis/Calgene_LLC

Figure 2.3.   Plant researcher Athanasios Theologis, showing GM tomatoes genetically modified to ripen, and hence spoil, more slowly. Image source: Wikimedia. Image credit: Jack Dykinga.

Only three years after the FDA approved the sale of Flavr Savr, Monsanto finished purchasing Calgene in 1997 [7]. This was also ironically the last year that its GM tomato was ever sold. The bust of Flavr Savr was generally not attributed primarily to it being a GMO, but rather to practical issues with the tomato business that Calgene had not grasped firmly enough. Nonetheless, the GMO commercial genie was out of the bottle and there was no going back. Many other companies were on the GMO bandwagon at that point as well.

Figure 2.4.   The former Calgene Campus at Monsanto in Davis, California, USA.[xi] Image source: localwiki.org/davis. Image credit Jason Aller. Reproduced with permission.

Theologis imagined that the antisense GMO approach to inhibiting ripening could have widespread applicability in the area of what was called "fruit senescence":

"The use of antisense technology and overexpression of metabolizing enzymes such as ACC deaminase in controlling fruit ripening is only the first step toward controlling fruit senescence…The prospect arises that inhibition of ethylene production using reverse genetics may be a general method for preventing senescence in a variety of fruits and vegetables."

This quote is important because some have suggested that GM humans could be produced to have inhibited "senescence," such that they would resist aging the way that GM plants have slowed ripening. The expression of people "living to a ripe old age" takes on a new meaning in the GMO era.

[xi] https://localwiki.org/davis/Calgene_LLC

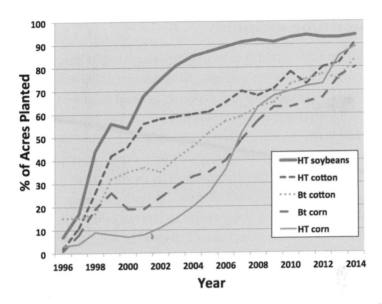

Figure 2.5. US Department of Agriculture data (used with permission) re-graphed by the author showing the spike in cultivation of GM cotton, soybean, and corn crops in America.[xii] HT stands for "herbicide tolerant" GM crops, such as those resistant to glyphosate (Roundup). Bt stands for insect-resistant GM crops with a toxin gene from the soil bacterium *Bacillus thuringiensis*.

At the same time that the GM tomato was in its infancy, Monsanto was well into another GM plant strategy based on an herbicidal chemical called glyphosate (better known commercially as Roundup). Researchers realized that they could make food crops resistant to glyphosate by using GMO technology. In this way, they reasoned, fields planted with GMO crops resistant to glyphosate could be sprayed with the herbicide, killing all the weeds, but leaving the food crops unharmed. This should and did boost crop yields. Glyphosate is now the most commonly used herbicide.

In a period of less than two decades, starting around the time of the Flavr Savr tomato, the percentage of certain key plant crops in the US that were GM crops jumped from essentially 0% to about 80–90% (Figure 2.5), sparking controversy discussed more in Chapter 8.

[xii] http://1.usa.gov/1CwvLSS

Democratizing creation

Over the years GMO-enabling technologies such as genomics, reproductive biology, stem cells, and gene editing have advanced to a very sophisticated level. Could this knowledge and expertise be used to make GMO sapiens? I do not see a clear technological obstacle to trying, but it would be risky.

We know the human genome sequence, we know how to edit this sequence with relative ease and precision (although improvements are always in the works), we can do these genetic modifications in a wide variety of organisms (including close relatives of humans, such as non-human primates), and these technologies work in human cells in the laboratory.

The final two ingredients are emerging as well: a potential market for GM people and increased public acceptance that GM people might be appropriate under certain conditions. To envision the future of GM animals and potentially GMO sapiens, we can look to Cambrian Genomics. Cambrian has talked about making designer babies at some future point, but the more immediate plan was to offer customers the opportunity to make designer GM animals of various kinds.[xiii]

Before his recent death, Cambrian CEO Austen Heinz (see interview in Chapter 5) pushed the boundaries of biotechnology. Heinz imagined a world in which GM animals and people can easily be made. Heinz and Cambrian had a mission to "democratize creation," meaning to let the public dive headlong without particular monetary or regulatory restrictions into the world of making GMOs. According to a recent newspaper article, Heinz had the goal of enabling others to make GMOs for just about any reason:

> "Anyone in the world that has a few dollars can make a creature, and that changes the game," Heinz said. "And that creates a whole new world."

A reasonable question is whether democratization of creation via GMO technology is a wise thing to do. I am not convinced it is. Up until now, making GMOs was more or less restricted to trained scientists in industry or academia. Cambrian wanted to overturn that mostly closed system and pave the way to a new reality of do-it-yourself (DIY) GMO.

[xiii] http://bit.ly/1tJkIX9

Was all of this just bluster intended to generate attention and more capital beyond the already impressive $10 million that Cambrian raised as of 2014? Perhaps there was some element of hype with Cambrian, but its words reflected a powerful notion that whimsical GM animals and even possibly GM humans are not only possible, but also desirable.

Another question relates to the economics. Heinz mentioned it will cost "a few dollars" to create a custom creature, but in reality it would likely cost tens or hundreds of thousands of dollars.

In the same article on making designer animals, Marcy Darnovsky, executive director of the Center for Genetics and Society, a genetics watchdog group in Berkeley, was quoted as viewing Cambrian's philosophy as a concerning "techno-libertarianism." There was also a cavalier attitude by Cambrian about making GMOs. Do we really want people, willy-nilly, making GMOs? As much as there is a potential fun element to this (see more below), there are risks too. It is unclear if Cambrian will continue without Heinz.

GM pets and novelties

A number of commercial efforts are underway to create GMOs to sell as pets, conversation pieces, or novelties. Should these GMOs enter the wild and become established, which is at least hypothetically possible, they could interbreed with wild organisms and lead to the creation of new species that contain genetic modifications and mashups of traits.

Heinz of Cambrian Genomics was also involved with another company called Glowing Plants that has produced GM plants that literally glow. The first ever GMO pet, bioluminescent fish (see Figure 2.6), was produced and is sold by GloFish. Author Emily Anthes' fascinating book *Frankenstein's Cat*, which is focused on GM animals, devotes a whole chapter to GloFish [11]. Anthes went so far as to buy some as pets:

"I set up the tank in my living room. Under the bulb's blue light, the Glofish gleam like jewels…it's entrancing to watch them swimming around, a kaleidoscope in constant motion. These fish may be frivolous, but they're just a teaser, a preview of coming attractions. If we can get

Figure 2.6. GM fish produced by the company GloFish. The fish are a genetically modified form of zebrafish that fluoresce. Consumers can buy them as pets. Image source: www.glofish.com.

> black-and-white fish to glow neon red, green, and orange, what else can
> we get animal bodies to do?"

GM mosquitoes as good "weapons" to fight disease

To answer Anthes' question, it turns out that GM animals can likely do much more. For a more serious application, it has been proposed that a laboratory-engineered GM mosquito could be used as an environmental weapon to combat malaria-causing mosquitoes in the wild. The idea is two-fold: GM mosquitoes that would not cause malaria could be used to outcompete wild disease-carrying mosquitoes; or they could interfere with the reproduction of wild mosquitoes.

Research from the laboratory of Marcelo Jacobs-Lorena indicates that the malaria-resistant mosquitoes that they have generated do indeed have a fitness advantage and they happen to have cool and creepy green fluorescent eyes too (Figure 2.7) [12]. The predicted outcome if these GM mosquitos were released in areas where malaria is endemic, is a

Figure 2.7.   GM malaria-resistant mosquitoes that express green fluorescent protein (GFP) in their eyes. Image credit: Marcelo Jacobs-Lorena. Reproduced with permission.

reduction in human illness. However, these supercharged insects, in doing the bidding of humans, could have unintended consequences if released into the wild from the laboratory.[xiv] They could affect the overall ecosystem and impact other plants and animals besides mosquitoes.

If there were negative consequences once the GM mosquitoes were released, there is no clear plan for reigning them in effectively. However, a limiting factor in the potential harm they could cause comes from the fact that mosquitoes are very restricted in where they can live. Everyone hates mosquitoes that spread malaria and other diseases (e.g. mosquito-transmitted West Nile virus is common here in Davis) so they are easy targets, but scientists are carefully thinking through whether and how they might be used. Others have proposed similar strategies, such as GM ticks, to control Lyme disease.

Another more powerful approach to tackling disease-carrying organisms than simple competition is something called "gene drive." In

[xiv] http://bit.ly/1cCyDqF

gene drive, scientists would intentionally try to force genetic change. In this approach, GM mosquitoes, or other GM insects, are released into the wild to rapidly genetically modify the whole existing population in a way that either wipes them out or reduces their ability to cause disease. See Chapter 7 for more on the exciting and frightening concept of gene drive through GMOs.

Defining human genetic modification

A genetic modification is the change of a specific genetic element from one state (e.g. the starting DNA sequence) to another. This could range from just a single A, C, G, or T alteration in DNA all the way up to changing an entire gene or even going as far as modifying a larger chromosomal region. Another possible change is insertion of a foreign DNA element such as the colorful genes in fish or mosquitoes.

Oddly enough when it comes to humans, some bend the rules as to what they assert would constitute a genetic modification. For example, as will be discussed more in Chapter 4, some argue that genetic modifications that try to prevent mitochondrial diseases are not really true genetic modifications. They claim that combining the nucleus of a mother-to-be's egg with an entirely new repertoire of mitochondria and hence changing 37 entire mitochondrial genes, should not be called "genetic modification." However, puzzlingly, changing a single nucleotide — say from an A to a T in a *nuclear* genome — would be considered a genetic modification in their way of thinking.

In the case of the Flavr Savr tomato, which is established as a GMO, a naturally occurring tomato gene was simply flipped around to code an inhibitory "anti-sense" sequence. That was it. The argument for why swapping out all mitochondria, and the more than 16,000 DNA base pairs per mitochondrial genome, should in contrast supposedly not be considered a genetic modification is highly questionable and may be more political than scientific.

For purposes of clarity and consistency, in this book we will refer to a heritable, specific change in DNA sequence in an organism to be a "genetic modification" and hence that organism, whether it is a plant, bacterium, human, or say a moose, is a GMO. In contrast, normal human or other organismal breeding as discussed further below does not produce what I will call genetic modification.

Some proponents of making GM humans, seek to minimize or normalize the possible human genetic modification process in various ways. One of the most common strategies is to compare human genetic modification to how human sexual choices (i.e. decisions about our own breeding) have been controlled at the individual, societal, or cultural levels.

They argue that even at the simplest level of choosing a sexual or reproductive partner, human beings are in a sense engaging in genetic modification of their future offspring. In other words, they claim that every human reproductive decision is an act of human genetic modification.

While it is accurate to say that every human couple that produces a baby has produced a human being with a novel genome (except in the case of identical twins) that is genetically different from that of the parents, our normal reproductive process is not at all equivalent to a laboratory-based effort to create designer human beings with specific genetic modifications. Every child normally has a random 50–50 mashup of their parents' genomes that is simply a result of normal sexual reproduction. A GMO sapiens baby would have a specific gene altered by design in defined ways in a laboratory. Normal human breeding is, by analogy, like shuffling two decks of cards together and then randomly pulling out 52 to form a new deck (the baby). Genetic modification, by contrast in that analogy, would be stacking the deck by reordering specific cards manually in a very specific way that you think will be favorable to form the new 52-card deck.

Humans are not the only living things that can genetically modify other organisms. Certain kinds of bacteria have a proven capacity to genetically modify plants (for example, by insertion of foreign genes) in a heritable fashion. For instance, Agrobacterium bacteria regularly modify the genomes of plants including sweet potatoes.[xv] These changes can alter the plant's traits and also in some cases cause galls, tumor-like growths, to

---

[xv] http://www.nature.com/articles/nplants201577

form. Viruses can also genetically modify the organisms, including the people, that they infect. All of us carry genetic modifications in some cells from viruses and these sometimes can cause diseases such as cancer.

Gene therapy

Another type of human genetic modification that is already taking place is "gene therapy," which only modifies certain kinds of cells in a human patient. It is specifically not intended to work in reproductive cells and steps are taken to avoid this possible outcome. For this reason, gene therapy does not lead to heritable changes in people. I will not in this book consider patients who have had gene therapy to be GMO sapiens. Gene therapy has its own set of scientific and ethical questions.

How does gene therapy work? The intent with gene therapy is to correct, in a single individual, a genetic defect that is associated with a disease, such as a blood disorder or cancer. To achieve this goal, somatic (non-stem and non-reproductive cells) cells are typically infected with a viral vector, which transmits the "gene correction" to the diseased cells.

These GM cells are either already present in the patient or, if grown in a laboratory, are then transplanted into the patient suffering from the disease. As a result, the patient essentially becomes a chimera. Most of the cells in their body retain a disease-causing mutation, but specific critical cells (e.g. blood cells) possess the key genetic modification that is hoped to confer clinically significant resistance to the disease in question.

The best example of a disease being targeted with gene therapy is severe combined immunodeficiency, also known as SCID. Boys or girls with SCID have to be in isolation in a sterile environment (they are often placed in protective bubbles and so sometimes the colloquial name "bubble boy disease" is used) because they do not have a functional immune system and so are very prone to infections, which can easily become life threatening.

There are a number of genetic causes for SCID. In X-linked SCID or "X-SCID," patients have mutations in a gene called "common gamma chain" located on the X chromosome. As a result, X-SCID patients lack many forms of a type of growth factor called interleukins that are vital for normal immune (white blood) cell development and function.

In another form of SCID, patients have a mutation in the gene encoding adenosine deaminase (ADA), an enzyme that breaks down a class of molecules called purines. Lacking ADA, purines accumulate in these patients, leading to a profoundly weak immune system and even to death.

In these two examples of SCID, patients have inactivating or "loss of function" mutations. Therefore, gene therapy can in theory treat or cure such patients by introducing a compensatory functioning gene to replace the missing, needed molecule. These genes can be reintroduced into the patient's blood cells and in principle these cells should make the patient healthier overall.

The main non-gene-therapy approach to treating SCID is bone marrow transplantation, after the malfunctioning immune system has been destroyed by chemotherapy. But this therapeutic approach does not always work very well overall for SCID. Therefore, there has been great hope that gene therapy might considerably improve the treatment of SCID. Recent articles and reports (at the time of writing) have produced very encouraging results on using gene therapy to treat SCID, which suggest it is superior to certain types of marrow transplants.[xvi,xvii]

Beyond blood disorders, other kinds of gene therapy are more problematic or impossible from a technical perspective. For example, not all organs or tissue are amenable to gene therapy. While the immune and blood systems are uniquely suitable to gene therapy in patients with genetic mutations that cause blood disorders, treating other tissues via gene therapy is far more problematic.

The key factor is that the diseased immune system and blood-forming cells can be effectively removed in total via chemotherapy. Then they are replaced. In genetic diseases that affect other tissues, such as the brain or muscle, such removal of mutant cells and diseased tissue is impossible, presenting a major roadblock.

[xvi] http://bit.ly/1F6sayS
[xvii] http://bit.ly/1NZSM86

Still, in a broader sense, gene therapy is an individualized form of non-heritable genetic modification that provides real hope for many patients. Scientists are working to reduce the associated risks. The focus of this book is instead primarily, but not entirely, on heritable genetic modifications in humans.

References

1. Berg, P, *et al.* (1975) Summary statement of the Asilomar conference on recombinant DNA molecules. *Proc Natl Acad Sci U S A.* **72**:6:1981–4.
2. Berg, P, *et al.* (1975) Asilomar conference on recombinant DNA molecules. *Science.* **188**:4192:991–4.
3. Kevles, DJ (1994) Ananda Chakrabarty wins a patent: biotechnology, law, and society, 1972–1980. *Hist Stud Phys Biol Sci.* **25**:Pt 1:111–35.
4. Sun, M (1986) Calgene breaks new ground. *Science.* **231**:4744:1361.
5. Theologis, A, *et al.* (1992) Modification of fruit ripening by suppressing gene expression. *Plant Physiol.* **100**:2:549–51.
6. Meyer, R (1995) Detection of genetically engineered plants by polymerase chain reaction (PCR) using the FLAVR SAVR tomato as an example. *Z Lebensm Unters Forsch.* **201**:6:583–6.
7. Brower, V (1997) Monsanto swallows Calgene whole. *Nat Biotechnol.* **15**:3:213.
8. Seralini, GE, *et al.* (2012) Long term toxicity of a Roundup herbicide and a Roundup-tolerant genetically modified maize. *Food Chem Toxicol.* **50**:11:4221–31.
9. Powles, SB (2008) Evolved glyphosate-resistant weeds around the world: lessons to be learnt. *Pest Manag Sci.* **64**:4:360–5.
10. Xu, Q, *et al.* (2013) Family-specific differences in growth rate and hepatic gene expression in juvenile triploid growth hormone (GH) transgenic Atlantic salmon (Salmo salar). *Comp Biochem Physiol Part D Genomics Proteomics.* **8**:4:317–33.
11. Anthes, E *Frankenstein's Cat: Cuddling Up to Biotech's Brave New Beasts.* First edition. ed.
12. Marrelli, MT, *et al.* (2007) Transgenic malaria-resistant mosquitoes have a fitness advantage when feeding on Plasmodium-infected blood. *Proc Natl Acad Sci U S A.* **104**:13:5580–3.

Chapter 3

Human Cloning

"Mightn't even you, in your heart of hearts, quite like to be cloned?"

— Richard Dawkins, author of *The Selfish Gene*

"Can you please clone my deceased child?"

— half a dozen people who have emailed me

A student and the first clone

In his teens, student John Gurdon struggled in science. He bombed at tests and at least once finished last in his class. One of his teachers wrote on his report card, "I believe he has ideas of becoming a Scientist; on his present showing this is quite ridiculous."[i] He forged on. Only about ten years later, now graduate student Gurdon was assigned a tough, but exciting project by his mentor at Oxford, Michael Fischberg. Gurdon was to clone animals, specifically frogs, to address the question of what happens to genes during development. His more specific task was to conduct nuclear transfer by moving the nucleus of a regular, differentiated cell into an egg that had its own nucleus removed. Would the new hybrid egg develop normally into a clone? Previous work by prestigious senior professors was not particularly encouraging. Earlier, the other scientists that had also tried to clone frogs had glimpses that it might work, but it

---

[i] http://bit.ly/1AH2rgM

had ultimately failed for them. Ultimately, Gurdon said he even surprised himself with his demonstration that he could clone animals [3].[ii] The world would never quite be the same.

Cloning culture

The idea of cloning has a powerful impact on people. It either excites or scares them. They start invoking the devil or the idea of playing God.

This array of complicated and often contrary feelings can trace its origins perhaps in part back to the idea of a doppelgänger ("double walker" in German) or more commonly referred to as "the double" in English. Someone once said to me, "I saw your double at the mall, Paul, and it freaked me out!"

Double mythology goes back thousands of years and was generally associated with bad things happening or with evil. Sometimes a person's double would be akin to an evil twin. There is also the unsettling idea that a double could take your place, possibly kill you off and assume your life. Dostoevsky's classic short story, *The Double*, unfolds in that vein. Even today in art and literature, cloning and the idea of doubles often make appearances.

On the more positive or tempting side for people, human cloning is also linked to a sense of immortality. The idea that a genetically identical form of us could be made and persist beyond our lifetime is, to some people, very attractive.

In the real world, cloning of a normal sort can be seen in nature, even in humans, in the form of identical twins produced during the same pregnancy. This occurs when a single cell of an early human embryo separates off from the rest (hence being genetically identical) and goes on to form an entirely distinct embryo and ultimately a person that is genetically the same.

With sometimes visceral associations as a backdrop, it was perhaps not too surprising that as biologists began to work on experimental animal cloning, it was viewed as both exciting and frightening by some fellow scientists and members of the public.

[ii] https://www.youtube.com/watch?v=TLBnidXc7Q8

Keep in mind that, as will be discussed later in this chapter, cloning can also refer to making cell lines that are genetically identical to people, so cloning does not always involve making a new person. The cloning of ESCs, for example, is called "therapeutic cloning," while the cloning of actual animals is referred to as "reproductive cloning."

The birth of cloning

Asexual cloning in the world of botany and agriculture has been a long-time observation and practice. Plants can create clones of themselves naturally via budding or underground sprouting of new cloned individual copies. For instance, strawberry plants can create identical clones of themselves, called plantlets, on their runner stems.[iii]

Some animals also perform a type of natural cloning via a process called parthenogenesis, where a female animal can produce offspring without a male. In parthenogenesis, an egg starts dividing even without having been fertilized and in some cases can form a healthy female offspring without the involvement of sperm.

Parthenogenesis is not known to naturally occur in humans, but can be triggered to occur in a laboratory with an artificial stimulus, such as an electric shock to the egg. Parthenogenesis has sometimes been used to make early human embryos and then in turn human ESCs independent of IVF. Otherwise, naturally occurring parthenogenesis is limited to only a few species, including some lizards, insects, and fish, to name just three.

Scientists studying cells had also observed that individual cells often essentially clone themselves every time they divide to produce two identical, or nearly identical, cells. However, as it turns out, sometimes cells divide to produce two so-called "daughter cells" that are very different from each other. In some stem cells, this process is called asymmetric division.

We all have two kinds of cells: somatic cells (everyday ordinary cells) and reproductive cells (also confusingly known as germ cells). Germ cells can give rise, via fertilization or through parthenogenesis as mentioned above for eggs, to new organisms. Somatic cells normally cannot.

[iii] http://bbc.in/1liVdqv

To clone an animal, a scientist starts with a somatic cell such as a skin or blood cell from the animal to be cloned, and also an egg from a donor. The scientist removes the egg's own nucleus and transfers in a nucleus from the somatic cell or in some cases the entire somatic cell. With a tiny shock, the new incarnation of the egg with the now somatic nucleus instead of its own will sometimes go down a parthenogenesis-like route and start growing into an embryo. If implanted, on occasion the embryo will make it all the way through development and create a cloned animal.

For example, if I wanted to clone myself, I could take one of my own everyday skin or blood cells and try to clone that. To do so, I would need a human egg. I would remove the egg's nucleus (which contains all of its own DNA coming from the woman who donated it). Then I would put my skin or blood cell's nucleus or the entire cell into that so-called "enucleated" human egg. The nuclear switch is referred to as "nuclear transfer" and the process is more generally called "somatic cell nuclear transfer" or SCNT.

The hope would be that with a little shock this new hybrid egg, containing only my nuclear DNA, could then start growing and go on to form a human being that would be my clone. I would need to find a surrogate to bear the child and pay her for the service. Of course I would never do this, but have just used this as an example of the process.

During SCNT, the mitochondria of the donated egg are retained and have their own small genome. As a result, clones are also genetically modified since they have the nuclear DNA of one animal to be cloned and the mitochondrial DNA of another in the form of the egg donor. Hence, clones are not quite 100% genetically identical to the original animal. Still they are nearly identical.

Early developmental biologists, who studied animal development during embryogenesis and adulthood, were of course not aware of what we know today of animal cloning discussed above, but still they made some surprising discoveries that paved the way for cloning. For example, they could remove the nucleus from an oocyte without destroying the remainder of the oocyte, potentially paving the way for SCNT. This was helped along by the fact that the oocyte of many species is often a relatively large cell compared to most somatic cells. Oocytes are around one millimeter in frogs, for example, and are visible to the naked eye. They have a correspondingly giant nucleus that is, in a sense, a big, easy target to remove. Human eggs and their nuclei are not that oversized, but still are really big compared to the average somatic cell.

Scientists found that they could also remove the relatively small nuclei from somatic cells and sometimes even do so with the nucleus remaining intact, even if the residual cell itself was left in pieces. In other words, they could take cells apart and occasionally not destroy all the separate parts in the process. But could they mix and match parts from different cells as they tried to put things back together?

At some point, a scientist for the first time must have had an idea along the lines of, "If I can remove an egg nucleus without destroying the rest of the egg, and if I can remove a somatic cell's nucleus intact, perhaps I can replace the egg's nucleus with that of the somatic cell and create one new, hybrid egg that might grow into a whole new organism."

Sounds simple, but it took a lot of work. Actually, cloning at least in part came about originally as a result of an attempt to tackle a different, but related and equally important question: how does a germ cell and its nucleus ultimately turn into all the many billions of different, specialized somatic cell types of mature organisms?

How was such a total gene function changeup such as that possible?

Was the DNA changed or was there just a change in how the DNA functioned?

Thinking along these lines and puzzling out such questions, two decades before Gurdon's work German embryologist Hans Spemann (Figure 3.1) in the 1930s wondered what would happen if he took a differentiated cell nucleus and put it in a germ cell. What would be the result? And could this provide insight into how germ cells and their genome can functionally change into differentiated cells with entirely different genes activated (or "expressed" as we biologists call it) during development? In a historical piece on cloning, the journal *Proceedings of the National Academy of Sciences* wrote about this work:

"Spemann reasoned that if an egg implanted with a nucleus from a differentiated cell still developed into a normal embryo, this would

Figure 3.1. Hans Spemann. Image source: Wikimedia Commons, photographer unknown.[iv] 1935.

prove that the nucleus retains a full genome capable of directing all types of differentiation. In other words, a differentiated nucleus could still be totipotent."[v]

A totipotent nucleus (once inside a cell) is a nucleus that is able to direct its cell to form any and all types of an organism's cells, including the extraembryonic tissues of the placenta and umbilical cord, for example. Spemann conducted some preliminary cloning type experiments and found that, after a few cell divisions in an embryo, the nuclei of those early embryo cells retained the ability to differentiate after SCNT.

[iv] http://commons.wikimedia.org/wiki/File:Hans_Spemann_nobel.jpg
[v] http://www.pnas.org/site/classics/classics4.xhtml

He was able to clone a salamander via this form of nuclear transfer using the nucleus of a cell that had undergone four embryonic cell divisions transplanted into an enucleated fertilized egg. He appeared to be hot on the trail of cloning.

The big caveat was that early embryonic cells, even though not germ cells, are still far from differentiated. They have not yet acquired specialized gene expression or cellular functions, such as the jobs that skin, blood, brain, and other mature cells perform. Therefore, the question of cloning in a sense remained somewhat unanswered even if Spemann's results were supportive of the idea that a somatic nucleus could become totipotent if transferred into an egg. Furthermore, his experiments importantly suggested that the DNA was not changed (e.g. old genes were not removed and new ones were not added) during development, but rather it was how the DNA functioned that was altered. However, the work did not quite prove it.

Some years later and across the Atlantic, Philadelphia researcher Robert Briggs wanted to do experiments on nuclear transfer. His work would turn out to be very important for the cloning field and helped to explain what happens to the genome during development. As today's researchers can attest, much of the specifics of one's research and when one does the actual work depend on the funding that one gets. Briggs applied to the National Cancer Institute for a grant on nuclear transfer and it was soundly rejected. The reviewers described the proposal as, "a hare-brained scheme" [1].

Briggs persisted. He eventually teamed up with another scientist, Thomas King, to try to master nuclear transfer in frogs, specifically a leopard frog with jumbo-sized eggs that were available with an almost limitless supply. With these eggs and a couple years of work, Briggs and King reported in 1952 that they were able to clone frogs into new embryos and tadpoles [2]. The somatic cells they used were older than Spemann's, but still relatively young embryonic cells from blastocysts and not from adult frogs. Briggs and King ultimately decided after further experimentation that it might be impossible to clone actual adult somatic cells. It just would not work for them. This sort of frustrating experience in science happens to all of us at some point in our careers. Do we give up or forge on?

Figure 3.2.   The stuffed remains of the first mammalian clone, Dolly the sheep. Image source Wikimedia. Image credit: Toni Barros.[vi]

It turned out that it was not impossible to clone adults, but just very tricky. Within 10 years of Briggs and King's first paper, Gurdon reported the successful cloning of another species of frog mentioned at the beginning of the chapter. Amphibian cloning of adult animals worked! Gurdon received the Nobel Prize just a few years ago for this amazing achievement.

At the time of its report, this success shook things up in the field of developmental biology and a flood of additional research resulted. What about "higher" organisms such as mammals? Would that sort of cloning also be possible? It would prove doable, but extremely difficult at first.

Astonishingly it took British embryologist Ian Wilmut over four hundred attempts to clone sheep before he was able to successfully clone Dolly (Figure 3.2), the first ever cloned mammal [4]. This means that there were four hundred odd failures to get the one success.

This unfortunate "by-product" of cloning work is one reason why people get so nervous about attempts at human cloning. What would you do with your failed attempts at human reproductive cloning? I do not see

---

[vi] http://commons.wikimedia.org/wiki/File:Dolly_face_closeup.jpg

any ethical answer to that question. The same concern applies to attempts to make GMO sapiens.

Dolly's coming into this world freaked people out even more than Gurdon's cloned frogs. It was such an amazing thing that some were very skeptical at first that Dolly was even a real clone.[vii] Some skeptics speculated that the supposed somatic cell Wilmut had used to make Dolly from the udder of a female sheep (which incidentally is reportedly the sexist reason why he called her "Dolly" after Dolly Parton because he liked the actress' breasts[viii]), might have been a rogue, residual embryonic cell remaining in Dolly's mother.

Genetic tests proved that not to be the case. She really was a clone.

Although Dolly was difficult to make, with many developmental failures along the way, and even though she died young, the cloning of mammals has been optimized over the years since. Reproductive cloning of certain types of animals is now relatively routine, even if still somewhat technically challenging to do.

After the cloning of Dolly, the world went a bit clone crazy, particularly the media. The idea of human cloning was much closer to reality with Dolly around, alive and kicking and baaing.

Perhaps not surprisingly, Gurdon was asked about the possibility of human cloning. In one piece in the British press, he appeared open to the idea, at least as quoted:[ix]

> "I take the view that anything you can do to relieve suffering or improve human health will usually be widely accepted by the public — that is to say if cloning actually turned out to be solving some problems and was useful to people, I think it would be accepted."

To clarify things, I asked Professor Gurdon about this newspaper quote and about his feelings regarding the possibility of human cloning in this interview:[x]

[vii] http://nyti.ms/1KwAhVe
[viii] http://www.synapses.co.uk/science/cloneqa.html
[ix] http://dailym.ai/1EYQcKU
[x] http://bit.ly/1F6sduB

Knoepfler: Assuming technological issues are overcome, do you see human cloning (and by this I mean reproductive cloning) becoming an accepted, relatively normal element of human society in say about 50 years?

Gurdon: I do not see any prospect of human (reproductive) cloning becoming an accepted element of society. This is because all cloning experiments generate a fairly large number of abnormalities, as well as some normal products.

Knoepfler: Would you be comfortable with human cloning? In part, I ask because this article[xi] seems to suggest that you would indeed be OK with the idea.

Gurdon: I would only be comfortable with human cloning if it could be shown that it works more efficiently than normal reproductive cloning.

Knoepfler: What do you see as the potential positives and negatives, such as certain ethical issues, of human cloning for society?

Gurdon: I see no significant positive outcome of human cloning unless, as I said above, it turned out to be so well developed that it gave a smaller proportion of abnormalities than normal reproduction.

I found this interview somewhat reassuring in terms of the level of caution expressed by Gurdon regarding human reproductive cloning, which was very different from the newspaper quote. My sense is that it is unlikely that human cloning would, for example, lead to lower levels of reproductive abnormalities than those that occur naturally in normal reproduction, which is Gurdon's hurdle. Even so, the possibility of reproductive human cloning is more imminent than ever today.

The two kinds of cloning

In these discussions it is important to keep in mind again that there are two kinds of cloning of animals, including humans. The first type of human cloning is the reproductive "Star Wars"-type that would be used to make a new human being that is genetically identical to another human. Second, there is "therapeutic cloning," which uses SCNT to

---

[xi] http://bitly/1F6sduB

make human ESCs that are genetically identical to an existing person, such as an adult. The word "therapeutic" is included because the cloned human ESCs are intended to be used as the basis for autologous (self-to-self-transplant) cellular medicine. At this point, it is unclear if cloned human ESCs will ultimately be used in medicine since ESCs made from leftover IVF embryos and other powerful stem cells called IPSCs are far ahead in the clinical pipeline. The two different kinds of cloning are outlined in Figure 3.3. Many stem cell scientists support therapeutic cloning, but strongly oppose reproductive human cloning.

To illustrate the difference, let us say I want to do both types of cloning on myself. Yes, scientists are not supposed to experiment on themselves, but this is just a hypothetical example. Via reproductive cloning, I could generate a "mini me" baby version of myself that is genetically identical to me as mentioned earlier in the chapter. Through therapeutic cloning, I could make ESCs that are genetically identical to me. I could in turn use those stem cells to make any number of differentiated tissues (e.g. blood, liver, kidney) that could then be used as needed medically as the basis for

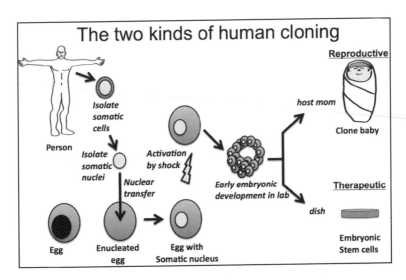

Figure 3.3. The two types of human cloning: reproductive and therapeutic. Image source: *Stem Cells: An Insider's Guide* [5] by the author. Often today instead of isolating the somatic nucleus, researches fuse the entire somatic cell with the enucleated egg.

a self-transplant with no immunosuppression. Cloning in either form is powerful technology. It could well be that cloned humans, even if apparently created successfully, may have any number of health problems and die young. We just do not know what would happen.

When the whole media storm happened after the cloning of Dolly, there were attempts to make human reproductive cloning illegal in the US, but conservative political leaders also wanted to outlaw therapeutic cloning. It appeared impossible at that time politically to only bar reproductive cloning so there was a stalemate. As a result, we are left today in the US with the odd situation where human cloning of either kind at the federal level is legal. Keep in mind that some individual American states and dozens of countries around the world have active bans on reproductive cloning.

Phony cloning

As with anything that generates so much attention, cloning also sparked its share of strange events and ultimately two instances of what appeared to be fake human cloning: one therapeutic and one reproductive. These alleged instances of cloning, although ultimately shown likely to be bogus, nonetheless generated a huge international firestorm. At first, perhaps most people thought that both kinds of human cloning had been successfully achieved, but it was not so.

The first type of phony cloning fooled almost everyone initially. Korean stem cell scientist, Dr. Hwang Woo-Suk (pictured in Figure 3.4) claimed to be the first to have successfully performed therapeutic cloning to make human ESCs, as reported in two research articles that he published in *Science* in 2004–2005 [6, 7]. These claims turned out to be untrue, according to a subsequent report by Korean officials.[xii]

Hwang was later found to have engaged in research misconduct, leading to the retraction of these two cloning papers, and every indication was that his therapeutic cloning itself was not real. Other ethically

---

[xii] http://nyti.ms/1zP2V3G

Figure 3.4. South Korean Scientist Hwang Woo-Suk after his cloning controversy erupted. Image courtesy of Rex Shutterstock, reproduced with permission.

concerning behavior reportedly involved compelling female scientists in his group to donate their eggs for this research, embezzlement, and data fabrication. Some of the women compelled to give their eggs ended up in the hospital as well.[xiii]

An odd side note to Hwang's fraudulent therapeutic cloning work is that the US Patent Office in its great wisdom nonetheless recently granted Hwang a patent on human therapeutic cloning.[xiv] Strangely, apparently something need not be definitely real in order to patent it.

[xiii] http://bit.ly/1J2Pocr
[xiv] http://bit.ly/1bH83v3

At least one of Hwang's other later claims to fame, being the first scientist to ever reproductively clone a dog, appears by all accounts to have been genuine. Hwang did clone the first dog ever [8], which he named Snuppy, a portmanteau or mashup of SNU and puppy, the former acronym standing for Seoul National University, where some of the work was done.

The cloning of Snuppy was even more difficult than that of Dolly, despite technological developments in the intervening years. Snuppy was the only surviving cloned dog from more than 1,000 implanted embryos in 123 surrogate mothers.[xv] It is still not entirely understood why dog cloning was so much more difficult.

Almost a decade after the Korean institutional report proving that the therapeutic cloning reported by Hwang was faked, and following serious professional repercussions for Hwang himself, he has now made something of a comeback, in addition to his recent patent success on therapeutic cloning.[xvi] He has his own company Sooam, which is in the animal reproductive cloning business. He is also involved in a collaboration in China that brings him together with therapeutic cloner, Shoukhrat Mitalipov,[xvii] whose advances in therapeutic cloning have been verified. I do not think we have heard the last of Hwang or his ties to animal and human cloning.

The other cloning fraud was perhaps, if you can believe it, even more outlandish than that committed by Hwang. This supposed human reproductive cloning was announced by a UFO cult called the Raëlians and their company, Clonaid. The cult is led by a former musician, Raël, who claimed to have had an extraterrestrial experience in which, amongst other things, he came away with the belief that cloning is a positive thing. Clonaid claimed to have cloned a baby girl, whom they called appropriately enough "Eve" in 2001. Clonaid CEO Brigitte Boisselier was a scientist who got drawn into the cult and ended up being their main spokesperson.

[xv] http://en.wikipedia.org/wiki/Snuppy#cite_note-riseandfall-3
[xvi] http://bit.ly/1zP2XbR
[xvii] http://www.ipscell.com/2015/02/mitalipov-hwang/

A court battle ensued as my friend, attorney Bernard Siegel, filed a suit related to ensuring the clone Eve's welfare, but there were suspicions that Eve was not real. Ultimately, Clonaid was unable to produce evidence related to Eve and claimed she had been taken out of the country so the US judge had no jurisdiction. When challenged to release DNA samples, the Raëlians then waffled and said that the "parents" of the clone wished no further contact, and there would be no DNA sample to verify the cloning.

Still in other press releases, as well as in statements by Boisselier, they went on to claim that they had cloned other babies as well. No baby Eve or any other Raëlian clone has ever been documented.

The claim by the Raëlians and Clonaid to have successfully performed human reproductive cloning still is widely dismissed as a hoax and publicity stunt.

The two apparently faked instances of human cloning more than a dozen years ago illustrate how cloning is such an explosive yet seductive issue and how news of cloning is irresistible to some in the media and even some scientists. It also suggests that future cloning hoaxes could occur again rather easily, particularly since today's technology has brought us much closer to human cloning really happening. The same level of media storm could also explode around faked human genetic modification.

Cloning myths

There are many misconceptions about cloning. These include myths and urban legends. These areas of confusion or outright bogus stories can muddy the water or be used to advance certain political agendas. One of the most widespread myths about cloning is that the clone created would be 100% genetically identical to the person intended to be copied. In reality because the cloning process involves an egg donor, the clone would be slightly distinct via having different mitochondrial as alluded to earlier.

More broadly, what are the most common myths?

Myth 1. Clones are just identical twins

Some folks like to argue that human clones would be no big deal because nature already makes clones in the form of identical twins, as we discussed earlier. Well, not really. These people have it all mixed up.

For example, commentator Richard Dawkins has frequently weighed in on human reproductive cloning, usually to downplay concerns:[xviii]

> "Anybody who objects to cloning on principle has to answer to all the identical twins in the world who might be insulted by the thought that there is something offensive about their very existence. Clones are simply identical twins."

That is not quite right.

While identical twins are clones, the reverse is not always true. Clones are not necessarily identical twins. Clones can also be laboratory creations made artificially and for specific motives that might not be entirely positive for humanity. Identical twins are also the same age, whereas a laboratory-made clone could never be the same age as the person who is being cloned. To put it another way overall, the context in which clones are made and some of their attributes differentiates them from identical twins in important ways, and therefore the effects of clones cannot be equated with the effects of identical twins.

Myth 2. All cloning is the same

As discussed earlier, there are two different types of cloning (Figure 3.3). Therapeutic cloning advocates have no general interest in reproductive cloning. Too often in the discussion over "cloning," the types of cloning are all muddled together.

Myth 3. Cloning is a ticket to immortality

Some people think that one way to become immortal is to keep being cloned — by having one's clones cloned. In reality, your clone would be a different person to you because they would have had a different birth

---

[xviii] http://news.bbc.co.uk/2/hi/science/nature/265878.stm

mother and therefore a different uterine environment during embryonic development. Your clone would also grow up being raised by different parental figures, eating distinct foods, breathing different air, drinking water from another source, doing a different job, and so on plus having different mitochondria.

Taken together, all of these environmental factors would change the clone in fundamental ways and hence they would not only be a different person than you, but also they might be dramatically distinct in their worldview, personality, and more. Self-awareness of their unique clone state and your motivation for making them could also dramatically change them and in potentially negative ways.

Myth 4. Human cloning is impossible

Technological advances have brought us to the point in 2015 where human reproductive cloning could well occur at any time in the next few years. It is not clear that there are any insurmountable technical obstacles to cloning a person today if potentially hundreds of failed attempts as well as the resulting problematic embryos and fetuses do not bother the would be cloners. That does not mean that it will work or that even if it does work, it would be easy. However, from my understanding of all the steps involved and the recent breakthrough in human therapeutic cloning by three different groups in 2013, including Mitalipov's team, I expect that cloning a person should be achievable. The key ingredients will most likely be an indifference to the harm that cloning could cause and an abundant supply of money.

Ironically, while we know today that the statement that "human cloning is impossible" is a myth, ten to twenty years ago people claimed that the statement "human cloning is possible" was a myth. The reality today is the exact opposite of what was claimed before.

Myth 5. Clones would look exactly the same

In many movies, clones look exactly the same as each other, and that idea of identical clone appearance is a common misconception. Important aspects of human appearance are non-genetic, with strong influences coming from the environment and even chance. While it is likely that clones would appear to be very similar as adults, they could well have some subtly different elements, such as height, weight, and pigmentation. Even identical twins who share the same prenatal and postnatal environments can have some clear differences in appearance. The popular TV show, *Orphan Black*, focuses on clones. It does not address the issue that the clones could well have noticeably different appearances, as the clones on this show often take on the identities of each other. However, in reality those around them might be able to tell the difference between them based on moles, freckles, or other distinctive features that are present despite their nearly identical genomes; although perhaps there is an assumption that the clones can wear makeup to mask such differences. *Orphan Black* is discussed more in Chapter 8, which focuses on cultural views on human genetic modification and cloning.

One of the bigger misconceptions in this area of clone appearance is that a clone would just instantly appear to be like the person who was cloned. A clone would have to go through a pregnancy and would be born a baby even if cloned from an 80-year-old woman.

Myth 6. Cloning is illegal in the US

"Do not worry about cloning," I have read or heard, "it's illegal in the US and essentially all other countries!" As it turns out, this statement is a myth.

There is no federal law, or even binding FDA regulation, that would make it illegal to clone a human in the US. There are specific state laws that ban cloning and/or any research that destroys embryos, but there are plenty of states that do not have such laws. As a result, someone intent upon cloning a human being could well do so right here in the US without a clear threat of legal repercussions. If attempts at cloning led to illness or even to the death of human cloned fetuses or babies, then a legal reaction could well be quick and severe. And if you picked the "wrong" state to do it in, you would probably end up in jail even if the cloned baby turned out OK.

As mentioned earlier, there have been discussions about making cloning illegal in the US, but again no federal law was passed. Although many countries do ban reproductive cloning, there are also other countries around the world where there is no formal legal prohibition on cloning including Honduras, Nepal, Bangladesh, and Ethiopia just to name four[xix] that much the same as the US have no prohibition.

Note that there is no country or state where cloning is explicitly permitted.

The politics of cloning

Despite the eventual consensus of fraud, the Raëlian claims were taken somewhat seriously at the time and, even prior to the 2001 claim, Boisselier achieved widespread attention even amongst scientists and politicians. She and the leader of the Raëlian cult, Raël, made many appearances (Figure 3.5). She was even invited to attend a panel discussion on human cloning hosted by the prestigious US National Academy of Sciences, leading to a firestorm of criticism [9].

The kooky and scary cloning claims of Clonaid and the Raëlians also spurred calls for legislative action by the US government and other countries to ban human cloning. At one hearing in the US, a bioethicist, Arthur Caplan, weighed in with this:[xx]

> "Arthur Caplan, head of the Center for Bioethics at the University of Pennsylvania, told the hearing that immediate safety concerns should supercede (sic) any ethical discussion. Such concerns are so great, he said, that only 'cults, cranks, kooks and capitalists' would propose cloning a human given the known risks."

The problem is that the world is full of cults, cranks, kooks, and capitalists. The same folks who are enamored with human cloning could well fall in love with the idea of human genetic modification as well and give it a try.

In the 1990s and 2000s, cloning was much more in the news and the issue became heavily politicized in the US and elsewhere. Most people, including politicians, were aghast at the idea of human clones, while others

---

xix http://bit.ly/1JXTfXY

xx http://www.nature.com/nature/journal/v410/n6829/full/410617a0.html

Figure 3.5.   Boisselier and Raël present at a policy meeting. Image source: Rex Shutterstock. Image credit Tony Kyriacou. Reproduced with permission.

supported cloning. Many countries have banned human cloning including those in the European Union, Australia, China, and Canada to name a few. Cloning remains legal in most US states and in many countries around the world. Someone with enough money and desire to try to make human clones could likely find a place to try it in the coming years.

De-extinction: resurrecting the extinct

The emerging trend of trying to bring extinct organisms — such as mammoths, passenger pigeons and even dinosaurs — back to life, while very enticing, could also be conceptually and culturally paving the way towards the production of GM people as well.

I am not convinced that these so-called "de-extinction" cloning projects will succeed and indeed I was, and still am, one of the most vocal, public critics of these efforts.[xxi] Still, de-extinction is popular with the

xxi http://bit.ly/1IpCe8E

public, with some environmentalists who mourn the loss of extinct species, and with some members of the transhumanist movement. De-extinction is admittedly a fun idea, but has many potential risks. These efforts also subtly increase the public's acceptance of the idea of cloning in general and, more specifically, of human cloning. Some have even proposed bringing Neanderthals back to life.[xxii]

The projects to bring Mammoths back from extinction bring us back to two other players with ties to potential human genetic modification: controversial Korean human cloning researcher Hwang Woo-Suk mentioned earlier in this chapter and Harvard geneticist George Church. These two scientists have wildly different backgrounds and track records (Church has a stellar scientific record beyond reproach), but nonetheless both want to resurrect mammoths. They are on separate teams trying to do this. Church also has proposed de-extinction of Neanderthals, a now extinct type of human.

As someone who has had their genome analyzed by 23andMe and found that I, like most people, have a tiny bit (a few percent) of Neanderthal DNA, I can attest to the fact that Neanderthals were really a type of human being. The Neanderthal cloning project would require women to serve as surrogate mothers. In his book *Regenesis* (on page 11, [10]), Church put it in this provocative way:

> "If society becomes comfortable with cloning and sees value in true human diversity, then the whole Neanderthal creature itself could be cloned by a surrogate mother chimp — or by an extremely adventurous female human."

That controversial last line was apparently more philosophical or hypothetical than a hint at any real plan, as Church was later quoted that he was not literally looking for an adventurous female human for this kind of experiment.[xxiii]

If successful, it remains unclear what the cloners would do with these new Neanderthal humans and the same holds true for the Mammoths.

---

xxii http://wrd.cm/1EJHWhe
xxiii http://bit.ly/1Ozw0Cr

Would the latter be doomed to be circus-like spectacles at theme parks? Could we end up treating cloned Neanderthals as a subclass of humans, forever changing society?

Cloning meets GMOs?

Could cloning be viewed as yet another type of human genetic modification? Could cloning be involved in the production of GMO sapiens?

My answer is "yes," in both cases.

Cloning-related technology could come into play for human genetic modification. During cloning, scientists might be able to efficiently generate new genetic forms of people that are identical to past humans, except with one or more very specific genetic changes introduced. One technical advantage of making GMO sapiens this way is that the team involved could try many times to make somatic cells with this desired genetic change; even if it did not work that well or efficiently, no harm would be done. Ultimately, they would just need one perfectly edited somatic cell line and you could then make a GM human via cloning using the nucleus of that GM somatic cell. Perhaps you could in theory make many copies.

At some point, human GMO experiments might "get it right" to produce what are viewed as superior human beings, whether that step involves cloning or not. In principle, if certain gene-editing techniques consistently produced "better" people, what we might call "reproductive engineers" may try to duplicate the same now-proven GMO method on other embryos. However, there might be pressures to take another route, which is to clone those already produced gene-edited human beings that turned out so well. If you made a GM baby with the desired traits and no problems surfaced, for example, why not just duplicate it by cloning?

Cloners and friends

Today some people, like Dawkins mentioned earlier, are advocates of the idea of human cloning. These cloning fans also include some scientists, lawyers, and ethicists.

Some fertility researchers have been on that bandwagon too. Both in IVF pioneer Robert Edwards' time and more recently, fertility doctors have been of the view that cloning might be used as a treatment for infertility. Edwards, in testifying before a UK regulatory committee, made this telling statement (note that "NT" stands for "Nuclear Transfer," the key cloning step where the nucleus of an ordinary cell is moved into an egg that has had its own nucleus removed):[xxiv]

"It is now done in rabbits and rats, NT goes to full term and normally, as far as I can see . . . It means that we can start to think of doing this instead of doing things such as collecting testicular sperm…You would simply get a drop of blood and use that instead. It probably brings the descent of the male a little further than his rather poor performance in my field because it would not be sperm. I do not know how it would go down! I think that a lot of people would get advantage from it. I am being 100% serious here. If all couples where there were no gametes in one of them could use this technique, it could be very helpful to a lot of people worldwide, an awful lot of people. I cannot be more definite than that."

Clearly Edwards was at least open to the idea of using cloning as a treatment of infertility. Another more recent example of human cloning for infertility comes from a fertility clinic that floated the idea of using human cloning to treat infertility.[xxv]

This clinic, The Center for Human Reproduction (CHR), published an article that discussed the work of Shoukhrat Mitalipov on the use of therapeutic cloning to make human ESCs. Remarkably, CHR suggested that this technology could potentially be applied to reproductive cloning to treat infertility:[xxvi]

"In a more controversial clinical application, this research, however, also potentially points towards the possibility of creating human

---

[xxiv] http://bit.ly/1DZxnTH
[xxv] http://bit.ly/1FWI8wh
[xxvi] http://bit.ly/1AOboz6

embryos for infertility treatments in this way. While current U.S. law prohibits human cloning for reproductive purposes, laws, can, of course, change over time, should the safety of this cloning process be established."

Notably, they got it wrong on their description of US law; as mentioned earlier to my knowledge, human cloning would not clearly be illegal in this country at a federal level and is legal in many US states.

Others view cloning as a good way to go if parents find themselves in the tragic situation of having had a child die. Why not replace the child with a clone? Over the years I have been asked numerous times if I could clone someone's child who had passed away. Is that really technically possible? Not in my laboratory, but possibly in certain specialized laboratories.

If the child to be cloned has a chronic, fatal illness and death can be predicted, cells could be isolated before death and banked in liquid nitrogen for use in subsequent cloning, or the cloning process could proceed even while the sick child is still alive. Alternatively, if the child dies suddenly, the cells could be isolated at that time.

Biologist Panos Zavos, who apparently now resides in the US, has claimed to have tried human cloning.[xxvii] For example, he announced in London in 2004 that he had transferred a cloned human embryo into a surrogate mother, but later there were no indications of a birth. Zavos, who some have compared to the Raëlians,[xxviii] worked with another fertility doctor, Severino Antinori, on cloning.

Pete Shanks, associated with the Center for Genetics and Society, has been following human cloning for more than a dozen years. He posted a report from 2001 on a meeting at which Antinori and Zavos were present to discuss human cloning.[xxix] Apparently another cloning advocate, Dr. Richard Seed, who has indicated his belief that cloning is a religious

---

[xxvii] http://www.zavos.org/library/UAAN_A_150346.pdf
[xxviii] http://dailym.ai/1zP3jyZ
[xxix] http://www.geneticsandsociety.org/article.php?id=2828

prerogative uniting man with God,[xxx] was also present. Shanks came away from the meeting very concerned:

> "It is now clear that Antinori, Zavos, and their colleagues could be disregarded were they not so dangerous. They are willing to accept risks that no one else considers remotely acceptable, in an attempt to do something that most people consider obscene. Yes, they have between them created a lot of babies. Yes, they may be able to make a clone, which might or might not be born without significant deformity. Certainly, they have the determination, probably the technical expertise, and undoubtedly the complete lack of conscience required to attempt such a thing."

As recently as 2009, Zavos claimed to have continued the reproductive cloning experiments with 11 more cloned human embryos transfers, but to my knowledge no evidence has emerged to support these claims.[xxxi] Zavos has a website that in part promotes the idea of human cloning.[xxxii] He has also published papers favoring the idea of cloning, including as a treatment for infertility [11, 12].

There are and will continue to be advocates of human cloning and even some willing to give it a try in the real world. The concern is that as technology continues to advance, these rogue efforts at cloning could accelerate and lead to harmful outcomes.

Who will be the first human clones?

If we are going to clone people, especially at first, only a very few people could be cloned. Who would warrant the distinct of being the first human clones? What would happen to them?

In their now classic article in the journal *Nature*, "Social Values and Research in Human Embryology" published in 1971 [13], Robert Edwards and David J. Sharpe, from George Washington University,

---

[xxx] http://en.wikipedia.org/wiki/Richard_Seed
[xxxi] http://ind.pn/1IpDOat
[xxxii] http://www.zavos.org/home.html

tackled these questions as well as others on cloning and human modification (see more in the next chapter from that article).

Edward and Sharpe also imagined that potential IVF type technology could enable human cloning:

> "Experimental work in animals has shown that the chromosomes or nucleus can be removed from an ovum and replaced with a nucleus taken from a cell of a donor. The embryo would then grow displaying characteristics of the donor rather than its own. Nuclear transfer can be done manipulatively in amphibian eggs, and by means of viruses to fuse a donor cell with a mouse egg. There is no need to stop at one egg-several or several hundred eggs could be 'cloned' in this manner and might all develop into identical or closely similar offspring."

In the above passage, they imagined a future in which humans might be cloned as well as modified. What they describe as "technically difficult" in 1971 is now, in 2015, likely to prove relatively more straightforward with the intervening technological advances. Still, cloning a human being even today is not going to be a piece of cake and would be a very risky proposition. Even so, some people probably will try it again in the future.

Edwards and Sharpe then discuss the potential benefits and risks of cloned or modified children in a way that resonates powerfully now in terms of a disturbing potential clone reality:

> "Cloned children would be closely similar to an older, earlier individual, that is the nuclear donor, and therefore they would have been deprived of what would be interpreted as one fundamental human right, the right to be different. The cloned person could be intelligent, motivated, highly adapted to a technical society, just like the successful donors. He would probably be given a good education, be physically strong, and very successful. Such offspring could be so well adjusted and distinguished that any decision not to produce them could be strongly criticized. But it is equally possible that a child composed by nuclear transfers would be a psychological misfit. He might become infatuated with the donor's background, and would be virtually unrelated to his familial parents. There would be a strong impulse to treat cloned children as laboratory subjects to be analysed repeatedly during their lifetime."

This discussion highlights just how risky human cloning or modification would be. These would be unique human experiments in which the outcomes could not be known until the human subjects were already created and they could be facing enormous health risks. The subjects of the experiment, the humans to be created, could not consent to be experimental subjects in advance because they would not exist at that time. Keep in mind that some of the same issues apply to IVF as well including the lack of consent of the child.

Ultimately, the authors could not advocate cloning at least at the time of writing their article:

> "In our opinion, the objections and doubts about cloning, let alone the current technical difficulties or the problem of who is worth copying, dismiss any consideration of using this method at present."

That question of "who is worth copying" is very provocative. Even today, there are likely to be very diverse opinions as to who would be worthy of being cloned, assuming it was technically possible (e.g. cells or nuclei being available).

Marie Curie? Stephen Hawking? Mother Teresa? Gandhi?

You?

We may find out in a few years who the first human clone is and they will essentially be the living embodiment of an experiment.

In the next chapter, I discuss the first human genetic modification experiments that led to production of a few dozen GMO sapiens. How did that happen? How are those GM people doing? What can we learn from that experience?

References

1. Di Berardino, MARG McKinnell (2001) Thomas J. King Jr. 1921–2000. *Differentiation*. **67**:3:59–62.
2. Briggs, RTJ King (1952) Transplantation of living nuclei from blastula cells into enucleated frogs' eggs. *Proc Natl Acad Sci U S A*. **38**:5:455–63.
3. Gurdon, JB (1962) Adult frogs derived from the nuclei of single somatic cells. *Dev Biol*. **4**:256–73.

4. Wilmut, I, *et al.* (1997) Viable offspring derived from fetal and adult mammalian cells. *Nature.* **385**:6619:810–3.
5. Knoepfler, P (2013) *Stem Cells: An Insider's Guide.* World Scientific Publishing.
6. Hwang, WS, *et al.* (2005) Patient-specific embryonic stem cells derived from human SCNT blastocysts. *Science.* **308**:5729:1777–83.
7. Hwang, WS, *et al.* (2004) Evidence of a pluripotent human embryonic stem cell line derived from a cloned blastocyst. *Science.* **303**:5664:1669–74.
8. Lee, JB, C Park C Seoul National University Investigation (2006) Molecular genetics: Verification that Snuppy is a clone. *Nature.* **440**:7081:E2–3.
9. Bonetta, L (2001) Academies called to task over human cloning debacle. *Nature.* **412**:6848:667.
10. Church, GME Regis (2012) *Regenesis : How Synthetic Biology will Reinvent Nature and Ourselves.* Basic Books.
11. Zavos, PMK Illmensee (2006) Possible therapy of male infertility by reproductive cloning: one cloned human 4-cell embryo. *Arch Androl.* **52**:4:243–54.
12. Zavos, PM (2003) Human reproductive cloning: the time is near. *Reprod Biomed Online.* **6**:4:397–8.
13. Edwards, RGDJ Sharpe (1971) Social values and research in human embryology. *Nature.* **231**:5298:87–91.

Chapter 4

Messing with Mother Nature: The First GMO Sapiens

"I not only think that we will tamper with Mother Nature, I think Mother wants us to."

— Willard Gaylin, opening credits of the genetic dystopian film *GATTACA* [1]

The birth of IVF and a devilish dilemma

On November 9, 1977 two British scientists did the biological equivalent of breaking the sound barrier. The human race would never be the same.

Robert Edwards and his research partner Patrick Steptoe, who were introduced earlier in this book, made a healthy human embryo in a dish in a laboratory. That embryo was then successfully used to create an actual person. In essence, they created the start of a human being outside of a person. In the thousands of years of human existence prior to that day, embryos had instead only been made inside women after sexual activity.

To accomplish this feat of making an embryo in a laboratory, Edwards and Steptoe cleverly tapped into the human reproductive process; they took control of the fertilization and early embryo development steps. After that, they implanted the embryo back into a woman and let the process continue.

They broke into the fundamental process of human life and hacked the system. They successfully figured out how to do human IVF thanks to extensive IVF studies in animals and through an admirable effort of sheer willpower and persistence against all odds. It was great science that changed the world.

How did they make this revolution possible?

First, Edwards had the revelation that the successful IVF work in animals could be translated to people. He decided to try to replicate this success with human IVF and embryo studies in the laboratory:[i]

> "It was during the 1950s that Edwards began to think that he might be able to extend work he had been doing preparing eggs for fertilisation in animal species to treat women with blocked Fallopian tubes. 'It struck me what we should be trying to do was pluck the egg from the ovary and fertilise it in the laboratory,' he recalled. "We could do this in animals increasingly ... this was the way to go in the human species."

It is one thing to have a great idea and to want to pursue it until it becomes a reality, but quite another to have the tools, the materials, and the drive to achieve this. Regarding unique materials, in particular one rare and precious item was needed. It was not something you could just pick up at the corner market in a carton.

Humans eggs.

You need women to provide human eggs, which can be a difficult, painful and ethically complicated process. Sperm, in contrast, are easy to obtain. Edwards teamed up with Steptoe. Fortunately, Steptoe (a practicing obstetrician) was able to obtain human eggs through his clinical practice via hormone treatments of patients and harvesting by laparoscopy.[ii] They had the building blocks that they needed. Ethical egg procurement involves the consent of the donor, no coercion, and no monetary payment for the eggs themselves. Obtaining human eggs for different purposes around the world is not always ethically done as we discussed in the previous chapter with Hwang.

With the tools he needed, Edwards began working with human sperm and eggs in the laboratory in earnest. From all those deadly serious

---

[i] http://bit.ly/1bH8t4Q
[ii] http://www.laskerfoundation.org/awards/2001_c_description_p.htm

and mandatory sex education classes we endured in school as teenagers, in which we were warned that pregnancy could result practically just from looking at each other, we might imagine that fertilization and pregnancy via mixing sperm and eggs in a dish would be a simple matter.

It was not.

Edwards and Steptoe thought that the key would be to follow the embryo to the critical point of five days after fertilization, a time after which the 100-cell human embryo (called a "blastocyst" at this stage, see examples of early human embryos and blastocysts in Figure 4.1) could be implanted into a woman to try to establish a pregnancy.

Very early human embryos

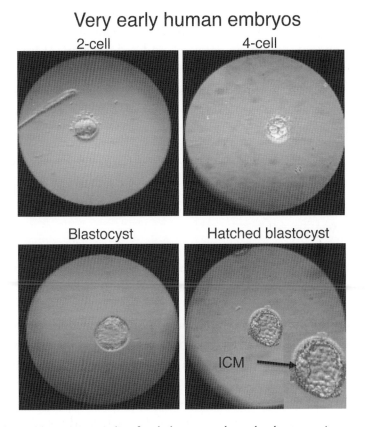

Figure 4.1.   Photomicrographs of early human embryo development. An arrow in the inset higher magnification view of the "hatched" blastocyst above, indicates the inner cell mass (ICM) that can produce the actual human body during a pregnancy or embryonic stem cells when cultured in a dish in a laboratory. Image credit: Meri Firpo. Reproduced with permission.

Could they make a human blastocyst in a dish? After much work, Edwards recalled an "aha" moment when they reached that goal:

> "I'll never forget the day I looked down the microscope and saw something funny in the cultures ... what I saw was a human blastocyst gazing up at me. I thought: 'We've done it.'"

Despite the fact that this was a major step forward in the laboratory, and notwithstanding the huge excitement of the researchers involved, this was still very far off from creating a healthy pregnancy and a new human being via IVF. To make matters worse, the reception outside of their laboratory to this important human reproductive research development was quite different to the reaction inside of the laboratory. Many people strongly opposed the human germ cell and embryo work of Edwards and Steptoe. Some of the harshest criticism came from other scientific colleagues, as well as from religious leaders. Pioneering DNA scientist James Watson even went so far as to suggest that Edwards might be "dabbling with infanticide."[iii]

It is unusual for scientists to so intensely attack other scientists in the public domain.

The Catholic Church violently opposed the work. The Pope condemned it. Steptoe and Edwards were acutely aware of the controversy as Steptoe was quoted:[iv]

> "I am not a wizard or a Frankenstein," he said. "All I want to do is to help women whose child-producing mechanism is slightly faulty."

Due to the public outcry, they faced practical roadblocks. In a major setback, Edwards was blocked from obtaining funding for the work inside the UK. At that point, even though things did not look good, they still forged on with the work.

Supported by an anonymous American philanthropist, Edwards and Steptoe secured the funding to continue their research and to attempt to

---

iii http://bit.ly/1bH8+4Q
iv http://articles.latimes.com/1988-03-23/news/mn-1905_1_patrick-steptoe

master the use of IVF to create a human being, a child for an infertile couple. And their research continued… and it continued; for five long years, their efforts did not work to create a human baby by IVF. In science, five years is an eternity. I can only imagine, as a fellow researcher, the intensity of their frustration. What could be the problem?

They tried many different adaptations of the technique. Eventually they shifted from focusing on blastocyst stage embryos with approximately 100 cells, and turned to a much earlier developmental time point — to embryos with only eight cells. This would turn out to be the breakthrough. At one point, they harvested a human egg from an infertile woman named Lesley Brown, mixed it in a dish with sperm to fertilize it, and were ultimately able to produce such an eight-cell embryo in their laboratory.

They then implanted that embryo back into the womb of Lesley Brown, who later gave birth to the first healthy "test-tube" baby, named Louise, in 1978. It was like splitting the atom in terms of scientific and societal importance, but in this case, instead of breaking something apart, they were putting something new together. A baby.

You can almost feel the combination of excitement, relief, and accomplishment in Edwards' and Steptoe's faces in the image from the press conference in 1978 (Figure 4.2).

A developmental phase of Louise Brown's existence took place outside of the human body; that was true of none of the other of the billions of humans who had come before her. For the first time in history, a human baby was born in part via a laboratory process, rather than from sexual activity.

Today it does not feel so unique, and more than five million babies have since been born to infertile couples via IVF, producing an immeasurable amount of joy and goodness. These millions of human beings would never have existed without Edwards and Steptoe, and many of their parents would never have become mothers and fathers without this innovative technology.

However, early on Edwards himself realized that IVF could be used in other ways too, and some of those were ethically far more complicated. He was thinking about the potential power and controversial applications of IVF seven years before the creation of Louise Brown, when he wrote the article, "Social Values and Research in Human Embryology" published

Figure 4.2.   Photograph of Robert Edwards (left) and Patrick Steptoe (right) in 1978 at the announcement of the first test-tube baby. Image source: Daily Mail/Rex Shutterstock. Image credit: John Sherboune. Reproduced with permission.

in 1971 [2]. Edwards teamed up to write this piece not with another scientist for a co-author, but rather with a lawyer from the US, David J. Sharpe, from George Washington University.

Why a lawyer and not another scientist? I imagine that Edwards was fully aware that IVF could not be viewed only in a scientific or even medical bubble. It would have profound societal and legal implications.

In the 1971 article, Edwards and Sharpe discussed how this technology, if realized in humans, could be a revolutionary treatment for infertility, which was essentially untreatable at that time. But they went much further. They also had the foresight to articulate how IVF hypothetically could be used to more selectively influence reproduction of human beings. For example, they wrote about how IVF could lead to eventual sex selection, with all of its complicated baggage as well. In that regard, they were quite right. IVF, together with preimplantation genetic diagnosis (PGD, see Glossary), is now frequently used for selection of female or male embryos.

Is such a sex selection use ethical? This question is still debated to this day and is discussed further in Chapter 5.

Furthermore, in a section of the article entitled "Modifying Embryos," they looked into a very accurate crystal ball and envisioned the current debate over the production or modification of human embryos in various ways. They wrote of two tantalizing, but risky possibilities. In this passage, they tackle the first:

> "Further developments that also depend on the growth of early human embryos perhaps contain the most controversial issues. These developments involve not simply identifying sex or genetic defects, but rather modifying or adding to the embryo itself. Two emerging possibilities are especially relevant to this time. First, donor cells can be injected into a mouse embryo and their multiplication during foetal growth can lead to the partial colonization of many organs. The foetus thus grows as a 'chimaera' with its own characteristics modified by those of the donor."

Therefore, in this first scenario, they raise concerns over the potential intentional future production of chimeric or hybrid humans who carry some cells of people who might not be their original biological parents. The second scenario they imagined was human cloning (more on that in the previous chapter). At the same time, they thought that this technology could have powerful medical applications for good as well.

Later, Edwards became more open with the potential power of his technological baby, IVF. It was the 25th birthday of Louise Brown, the first "test-tube" baby produced by IVF technology, when Edwards made the controversial statement that it was not God in charge, but rather the scientists in the laboratory.[v] Surely this claim by Edwards was a clear sign that IVF was about far more than just treating infertility. The point was that humans had wrested control of their own creation.

You can see Edwards with Louise Brown and the second individual to be created by IVF, Alastair Macdonald, in Figure 4.3, at a party celebrating IVF at the Bourne Hall Clinic in the UK, where some of the work was done.

---

[v] http://www.thetimes.co.uk/tto/life/article1716703.ece

Figure 4.3. Nobel Laureate Robert Edwards with the first two test-tubes babies, Louise Brown and Alastair Macdonald, attending "The Bourn Hall Baby Party" celebration of 25 years of IVF at Bourn Hall Fertility Clinic, Cambridge, UK on 26 July 2003. Image source: Rex Shutterstock. Reproduced with permission.

If scientists in a laboratory can be in charge of human creation, and even potentially manipulate it for different outcomes, then is this playing God or more simply humans rightly taking control of their own fate as a species? Do human beings with all their intellectual capabilities have the wisdom to be genetically tinkering with their own creation? I am skeptical that we are prepared to do so in a responsible, ethical manner even if technically we are ready.

A decade later, on the occasion of her 35th birthday, Louise Brown hailed Brown and Steptoe,[vi] and also emphasized that in her opinion IVF was not something to be concerned about:

> "When I was born they all said it shouldn't be done and that it was messing with God and nature but it worked and obviously it was meant to be…it is just the beginning of life that's a little bit different, the rest is just the same."

[vi] http://www.bbc.com/news/health-23448665

It makes sense that IVF babies once grown up would herald the inventors of IVF, but clones or GMO sapiens may not always feel that way.

Another, perhaps more ethically complicated side to Edwards himself exemplifies these kinds of dilemmas. Edwards was not just an innovative biomedical scientist, but also a believer in eugenics[vii,viii] and a leader in the British Eugenics Society. He was not a eugenicist in the negative eugenics sense of the Nazis or of the forced sterilizations that took place in the US [3], but rather in the positive eugenics sense of seeing an imperative to improve human beings through powerful technology. Should we be making such a distinction between the "positive" or "negative" forms of eugenics?

Edwards went beyond making human IVF a reality to envisioning its future use as a tool for "bettering" human beings [2]. Even so, from his words it is clear that he understood the risks along with the possible rewards of using IVF as a tool that opens the door to modifying embryos, in addition to its applications for treating infertility. In Edward's time, the risks were mostly hypothetical because the technology of the day could not clone or genetically modify humans but genetic engineering was racing forward in the 1970s too.

It is thought-provoking to speculate about whether Edwards' opinion would have changed if he were alive to see the technological advances of today such as CRISPR-based genetic modification of human embryos recently reported by the scientists in China. After reading much of his work, my impression is that Edwards would not be intimidated by today's opponents, including religious figures, and overall would likely support human genetic modification. Returning back to his famous quote about scientists rather than God being in charge, he had also remarked in response to opposition to IVF by the Pope at the time, "The Pope looked totally stupid. Now there are as many Roman Catholics coming for treatment as Protestants."[ix] In this case, by treatment Edwards meant IVF.

Beyond the ethical issues raised directly by IVF itself, IVF also gave birth to another very difficult problem: the creation of viable, sometimes surplus, embryos in the lab. This provided the opportunity to change them prior to their implantation. The complexity of outcomes and

---

vii http://www.scientificamerican.com/article/eugenic-legacy-nobel-ivf/
viii http://bit.ly/1F6sKN9
ix http://bit.ly/1bH8t4Q

applications of IVF technology raise a bioethical dilemma. While IVF can be used to treat infertility (a wonderful thing), it can also be used in a host of other ways, or combined with other technologies, that could have very different and even negative outcomes and raise serious ethical dilemmas, such as human sex selection, genetic modification, and cloning. Some object to the use of remaining IVF embryos for making human ESCs as well.

It is not that IVF was intended to be used for making GM people or clones. However, once a technology is developed and the information about the methodology is disseminated, others can run with it in new directions, and that has been true for IVF. A great example of the potentially negative part of this sort of dilemma in action was the very controversial infertility experiments described in the next section that relied upon IVF, but also included the modification of oocytes.

The "cowboys of medicine" make first GM babies

Gazing through a microscope in a laboratory in New Jersey, a scientist in Dr. Jacques Cohen's team (perhaps Cohen himself) used a special needle to jab into a donated human egg (see Figure 4.4 for an example of oocyte injection). The egg was sourced from a healthy, young woman. He then used a device to suck out part of that egg's cytoplasm (called "ooplasm" in eggs, short for oocyte cytoplasm), the gooey stuff inside of a cell that surrounds its nucleus and contains many important cellular components including mitochondria. He collected the ooplasm up into the needle.

Turning his attention to a second egg, he then inserted the same needle into it. This other egg is a donor from an older, infertile woman who wants to become a mother. The researcher then injects some of the younger woman's egg ooplasm into the older woman's egg in a process called "ooplasm transfer." The basic idea behind this was to make the older woman's egg healthier via an infusion of younger egg components and hence more likely to be successfully used to make a baby.

The resulting human egg now had cytoplasm from two different women. As a result, this new, chimeric egg, which contained mitochondrial DNA from the donor, was a GM human egg. The GM egg, once fertilized with sperm, was implanted into a uterus. In this case, this GM embryo

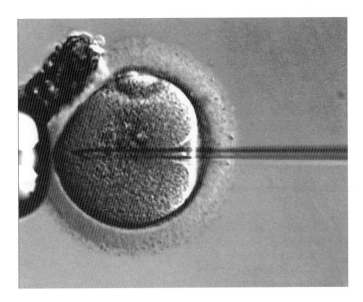

Figure 4.4. Photomicrograph of a human egg being injected with sperm using a very fine glass needle. The same kind of injection approach can be used to transfer the inner contents of a human egg to another egg in the technique of ooplasm transfer, a type of three-person IVF. Image source: Wikimedia. Image credit: Ekem.

grew into a living, human being. Thus, came into life the first ever GMO sapiens.

In this way, just two decades after the birth of IVF, Cohen literally punctured another barrier that had never been breached in human history. He was tinkering with the very substances that make us human and made a new form of human being.

This process was repeated with a series of eggs. Some grew into human fetuses after implantation and ultimately were born as additional GM babies. Notably, some other embryos made by IVF using chimeric eggs failed to develop. The technique did not always work, but it was amazing that in some cases it did. And so began a human experiment that built on IVF to create the first GM people, an experiment that in a way continues to this day as the GM children now grow into young adults.

Dr. Cohen and his former research partner Dr. James Grifo are reproductive specialists who pioneered this chimeric technique as well as numerous other revolutionary reproductive technologies. Cohen worked at The Institute for Reproductive Medicine and Science (IRMS) at St. Barnabas in New Jersey, a private fertility clinic and research center where he conducted the ooplasm transfer experiments.

At that time, as is also true to some extent today, the fertility industry was relatively loosely regulated in the US, especially compared to other countries such as the UK. Experimenting with human reproduction in these fertility clinics was not necessarily against the rules back then. Twenty odd years ago there was a push to try something fundamentally new to help infertile couples have babies as IVF alone did not always work, and Cohen and Grifo were among the pioneers, or from another perspective one might say the "cowboys," who forged ahead.

Some viewed this sort of work as being too far outside the norm. There were calls back then for greater regulation of the assisted reproduction field and some of those calls continue to this day. The fertility doctors did not view themselves as necessarily having freedom from regulatory oversight. For example, in a relatively recent interview with bioethicist Arthur Caplan,[x] Grifo recounted those days:

> "When I started 25 years ago, we were called the 'unregulated cowboys of medicine,' which was really not true. We were probably the most highly regulated area of medicine when you think about all of the regulators that we deal with."

Grifo and Cohen could be viewed as medical cowboys who were open to human experimentation in the name of trying to help people have babies. Their goal was to conquer infertility, whether by adding "young" ooplasm to older women's eggs or through other techniques.

Why mix together the components of two human eggs and how could that help infertility? This experiment was based on the premise that if the egg of an older, infertile woman had some suboptimal properties then transplanting some of the ooplasm from a younger, fertile woman's egg into the older egg might yield a healthier hybrid egg that could be

---

[x] http://www.medscape.com/viewarticle/817486 (note that you need to register for a free account to access this page).

fertilized by IVF and go on to make a baby. The younger woman's fertile egg was theorized to have some healthy "stuff" in it that would make the older woman's egg more functional.

Importantly, the donor egg cytoplasm has the mitochondrial DNA (mtDNA) of the woman who was its donor. This means that both the resulting embryo created by IVF from the hybrid egg and the subsequent person who grew from that embryo have the DNA of three people: a mother, a father, and a mtDNA donor. While mtDNA is relatively tiny compared to the DNA in the nucleus, keep in mind that it is still genetic material and has vital, delicately controlled functions.

Thus, this technique is a form of human genetic modification and it is heritable, meaning that all future descendants of these GM people will contain these mtDNA changes too. As a result, in a sense all descendants will also be GM people and have chimeric overall genetic states.

It was a radical, outside-the-box idea, and by all accounts it at least worked to the extent that some apparently healthy children were produced from previously infertile couples. That is a good thing, but not everything turned out quite so well. In total, one-to-three dozen (depending on accounts[xi]) GM children were produced by IRMS and other US clinics in this way before the US FDA stepped in to halt proceedings.

Why did the FDA put a stop to this technology in 2001?

We can see some of the reasons from the transcript of an FDA meeting at which Grifo was a participant and in which the work of Cohen was discussed.[xii] Dr. Deborah Hursh of the FDA voiced serious concerns:

> "In March of 2001, a laboratory of Dr. Jacques Cohen reported that two children born after the ooplasm transfer protocol were heteroplasmic, which means the genotypes of both the ooplasm donor and the mother could be detected in their tissues. These children were approximately one year old at the time of this analysis, so this was a persistent heteroplasmy that had been maintained.

[xi] http://humrep.oxfordjournals.org/content/16/3/513.full.pdf

[xii] http://bit.ly/1Az5J5v (PDF)

At the time of Dr. Cohen's publication the FDA was already considering action in the area of ooplasm transfer. The report of heteroplasmy raised our concerns, as did information in two pregnancies occurring after ooplasm transfer resulted in fetuses with Turner's syndrome, a condition where there is only one X chromosome."

What does this mean in regular English?

"Heteroplasmic" means, as mentioned earlier, that the fetuses made by these reproductive specialists had mitochondria from two different women, which could cause disease. The FDA also noted some serious chromosomal and developmental problems that occurred. Turner Syndrome, where females have only one X chromosome, comes with many serious health problems. Dr. Hursh further articulated at that meeting the FDA concern that the ooplasm transfer protocol may have been spreading rapidly to other fertility clinics. At that point 23 children had been born in the US from this technique.

Now that the children successfully produced are becoming adults, I wondered how they turned out. Are they healthy? Have any problems developed in them? Today, these questions are made all the more pressing by the push from some scientists to allow this type of experimental production of GM babies to resume for the specific purpose of preventing mitochondrial disease.

What do we know about the first batch of GM kids? Surprisingly, close to nothing:[xiii]

"Due to a lack of funding, Cohen says, it hasn't been possible to find out about how any of the children like Alana who were born from cytoplasmic transfer are doing. But the St Barnabus Institute is now starting a follow up study to check their progress."

The follow-up study only recently started and is being led by Dr. Serena Chen, a fertility specialist at the IRMS.

The "Alana" referred to in the article above is a teenage girl named Alana Saarinen, the poster child for the success of this method. She is by all

---

[xiii] http://www.bbc.com/news/magazine-28986843

accounts an entirely normal, healthy person even though she has DNA from her mom, her dad, and some mitochondrial DNA from another woman.

Alana herself appears very at ease with the situation:

> "A lot of people say I have facial features from my mum, my eyes look like my dad... I have some traits from them and my personality is the same too," says Alana.
>
> "I also have DNA from a third lady. But I wouldn't consider her a third parent, I just have some of her mitochondria."

Alana seems like a wonderful kid and it is impressive how comfortable she is with her unique situation.

Now some scientists are advocating for resuming other kinds of "three-parent" approaches that are similar in some respects to the old ooplasm transfer approach (more specifically to prevent mitochondrial disease rather than infertility). In the new method, instead of injecting some donor egg cytoplasm into the mom-to-be's eggs, the approach will be more involved because it will move an entire nucleus into a new egg. In the future version, the mom-to-be's nucleus (or the nuclear DNA itself) will be transferred into the healthy donor's egg. There will be an open space to receive this donated material due to that donor egg's own nucleus having been removed. The mom-to-be's remaining egg with the flawed mitochondria and no nucleus is then discarded. In this way, the idea backed by some reproductive scientists is to create a new hybrid egg for the mom-to-be that no longer has her mutant mitochondria.

These scientists argue today that the technique does not result in genetic modification and does not create GM people, but that is a tough case to make. I think they are wrong. Why?

First of all, the researchers who created the "three-parent" children such as Alana back in the 1990s and early 2000s discussed the genetics of the outcomes in no uncertain terms as genetic modification. In one of their articles [4], they conclude the abstract by saying, "This report is the first case of human germline genetic modification resulting in normal healthy children." Clearly, they believed the children that they helped to produce were genetically modified.

Even so, some have argued that three-person IVF experiments do not result in genetically modified people. The argument they make is based on three assertions that to me as a scientist are questionable: (1) the mitochondrial genome is so small that swapping it out does not really count as a genetic modification; (2) giving an embryo a new mitochondrial genome that is healthy in place of its mutated one is not really a "modification"; and (3) that because the mitochondrial genes themselves are not individually edited there is not a real modification going on here. Generally in science, changing even a single DNA base (whether nuclear or mitochondrial, wildtype or mutant) is defined as a genetic modification.

Although Alana appears to be doing great, other outcomes unfortunately were not so positive. In addition to the negative results cited by Dr. Hursh of the FDA, there was also a case in which one of the three-person IVF babies developed a "pervasive developmental disorder" (a series of multiple, very serious developmental problems including with cognition) in the first year of life, Dr. Cohen said. "Whether these anomalies were related to the procedure is unknown. The fact is that the parents could not become pregnant on their own or after conventional IVF. This could have also been the cause of the (Turner syndrome)," Dr. Cohen said. While this ambiguity does exist, and the parents' own genomes or other health issues could have been at fault, it is also possible that the ooplasm transfer itself and the resulting genetic modification caused the developmental problems.

Even though the FDA has blocked this work in the US since 2001, in 2003 another group, this time led by the American Grifo in China, did similar kinds of human GM experiments. Again, at that point, such work was banned in the US and still is to this day.

The experiment in China, according to *Nature News*, led to great controversy because the "Babies made by cloning-type technique die(d) prematurely."[xiv] Researchers at Sun Yat-Sen University in Guangzhou, China conducted this experiment in collaboration with New York scientists including Grifo.

[xiv] http://www.nature.com/news/2003/031014/full/news031013-4.html

The team produced human GM embryos by nuclear transfer in China.[xv] The three GM fetuses all ultimately died *in utero* at a gestational age range of 24–29 weeks. The researchers published their work and in the paper they stressed that the fetuses had apparently normal chromosomes.[xvi] The case of one mother is particularly tragic:[xvii]

"The 30-year-old Chinese woman underwent the IVF treatment at the Sun Yat-Sen University Hospital in Guangzhou, China, and became pregnant in 2003 with triplets after five 'three-parent' IVF embryos were transferred into her womb.

Doctors subsequently removed one of the foetuses to give the other two babies a better chance of survival but both these twins died after being born prematurely at 24 and 29 weeks respectively."

This outcome led to a fallout for Grifo, but he still defends the work:

"When Dr Grifo reported her case at a fertility conference later in 2003, he was widely denounced for carrying out an unsafe experiment of the 'three-parent' technique and was prohibited from carrying out further IVF trials…However, in an exclusive interview with. *The Independent*, Dr Grifo vigorously defended the work which he said demonstrated the viability and safety of a technique that could help women carrying inherited mitochondrial diseases to have healthy children of their own."

The work demonstrated "viability and safety"? It is hard to see how that is the case given the bad outcomes. The journal *Nature* cast the work in a skeptical light as an extreme, cloning-like technology:

"The new method comes close to human reproductive cloning, which is banned in many countries. In cloning, the nucleus of an adult cell, rather than of a fertilized egg, is injected into another egg so that the embryo is genetically identical to its parent. Grifo's technique creates embryos with genes from both mother and father. Like cloning, critics

---

[xv] http://www.wsj.com/articles/sB106600548112034100
[xvi] http://www.sciencedirect.com/science/article/pii/S0015028203019538
[xvii] http://ind.pn/1GaPeys

warn, Grifo's method might damage or incorrectly programme the mother's DNA. What's more, the embryos carry genetic material from two mothers: nuclear DNA from one, and small packages of DNA in the mitochondria from the other."

In the same interview with Art Caplan that was discussed earlier, Grifo was asked about emerging technologies. Some of his replies were quite striking and perhaps reflective of the very different perspective of the fertility industry on human modification. When queried about the future of infertility treatments, Grifo raised the issue, for example, of embryo selection, based on identifying DNA sequences that are associated with superior traits:

"With next-generation sequencing, we are learning more about genes. We are learning more about disease-causing genes. Should we select against that embryo? Should we not? Who should decide? Should the parents decide? They have to raise the child. I think they should decide, but we are going to learn a lot."

Should it all be left up to parental choice, even with highly experimental and potentially risky technologies? Encouragingly, when Caplan prompted Grifo with the idea of making "better" babies via new technologies, Grifo was disinclined to support that practice:

"'Better' is a funny word, because there has always been this criticism about designer babies, whatever they are. We don't know the genes for intelligence and hair color and eye color, so we can't select for those that easily. On top of that, you may not get the embryo that has all the things you want, and then what? Most of my patients don't care about that anyway; they just want to have a baby."

Even without pursuing the goal of a "better" baby, Grifo pushed the limits of technology. His three-parent IVF experiments raised ethical and legal questions that are still being asked to this day. Further, new genomics and genetics data are defining genes responsible for some traits such as eye and skin color as well as others.

Would it be legal to make a GMO sapiens?

Let us say I wanted to create a designer baby soon. Would it be legal? Could there be consequences for me, such as going to jail? Is it possible someone might sue me? The answers to such legal questions depend to a large extent on where one plans to produce the designer baby. In dozens of countries spanning the globe,[xviii] including most "developed" countries, it would be illegal [5].

For example, current law in the UK (with the exception of the three-person mitochondrial techniques) prohibits genetic modification of human embryos in a laboratory without approval from the UK regulatory body, the Human Fertilisation and Embryology Authority (HFEA). Therefore, the production of actual GMO sapiens via CRISPR, for example, would likely be prohibited by law there. Importantly, in many countries including the UK even sex selection of embryos by PGD is illegal.[xix]

In the US, things are not so clear, however, when it comes to designer babies and the law. While the FDA currently prohibits the creation of three-parent babies in the US, the legal situation regarding making babies with specific (for example CRISPR-Cas9-mediated) changes to their genomes is less clear. At present it would not be specifically prohibited at the federal level.

The BioPolicyWiki, which monitors international and specific individual national policies on human genetic modification, lists essentially no restrictions in the US[xx] and makes this blanket statement:

> "The United States has no federal laws or policies governing human genetic and reproductive technologies, although there are a number of such policies among the various states."

There are some federal policies, even if not laws. For instance, the FDA has indicated that three-parent IVF cannot proceed in the US at this time. Further, to be clear, while creating a designer baby is not prohibited

---

xviii http://bit.ly/1JXTfXY
xix http://bit.ly/1hzCUNs
xx http://bit.ly/1J13ysW

by US federal law, a long standing legal rule called the Dickey-Wicker amendment (D-W) prohibits NIH from funding any research that destroys an embryo [6]. American researchers who might want to make GMO sapiens cannot use NIH funding for such a purpose due to the legal restrictions of D-W since some embryos would not make it through the process. Even so, there are some very rich people interested in human genetic modification, and it would not take an outrageous amount of money to try to do it privately. In such a setting, D-W is unlikely to be much of a deterrent since its reach is limited to federally-funded research.

There are also some states in the US where production of designer babies would specifically not be permitted regardless of one's source of funding including Michigan, Minnesota, and Pennsylvania just to cite three.[xxi] However, these restrictions do not specifically prohibit the creation of designer babies, but rather would be invoked because of laws prohibiting research on embryos that leads to their destruction or that prohibit the creation of human clones or use of even therapeutic cloning.[xxii]

An American research team trying to create a GM person would have to practice on human embryos, and even during the "real" experimental production of a designer embryo, it would be unavoidable that at least a few embryos would be destroyed in the process. In principle, a team might try to get around this by not technically destroying either unneeded or damaged embryos by, for example, freezing them forever, but that would appear to violate the spirit of the state laws.

In China, officials have reportedly indicated that human genetic modification would need regulatory approval. As touched upon earlier, the first published attempts to do a cloning-like reproductive procedure to try to prevent mitochondrial disease came from work in China, leading to the reported, very negative outcomes. It was also where, in 2015, the first reported research on the genetic modification of human embryos using gene editing was conducted. The first attempts at making a GMO sapien via gene editing could happen in any one of many different countries including either China or the US.

[xxi] http://bit.ly/1LOP8vu

[xxii] http://www.ncsl.org/research/health/human-cloning-laws.aspx

Realistically, what might happen in the US if someone announced the intention to make GMO sapiens in the next few years? Or what if they just went ahead and tried without telling anyone first or asking for permission? Legal scholar and Stanford Professor Hank Greely, who was one of the authors of a commentary piece in *Science* calling for a prudent path forward on human germline modification, has some thoughts on what might transpire.

Professor Greely published a great post[xxiii] on "The Center for Law and Biosciences at Stanford" blog describing his views of human germline modification and its possible repercussions. He discussed a number of reasons why he is relatively less concerned about human germline genetic modification than say other applications of CRISPR technology, such as the editing of animals.

One deterrent to trying to make GMO sapiens that Greely noted was a potentially very serious legal liability: "If the moral risk isn't enough of a deterrent, the potential legal liability should be."

I replied with a post on my blog[xxiv] writing that I was not aware of any specific law against making GM humans. Greely then commented (excerpted below) on some specific examples of how the creators of a designer baby still might find themselves in legal hot water:

> "I think there are several venues for action against someone who tries, in our present state of ignorance, to make babies with germline genome editing. One is the FDA, where the jurisdiction isn't clear but a sympathetic case like this is likely to make it clearer, at least to the courts that review it. Another is civil liability, particularly if something goes wrong — either to the parents or to a damaged baby. Only a few states recognize wrongful life suits by babies for their prebirth damages, but, again, a sympathetic case like this…might make other state courts sympathetic. Third, consider miscellaneous criminal charges — the lack of a specific statute doesn't necessarily constrain a creative prosecutor. Child endangerment? Manslaughter (reckless or negligent killing) if a baby dies? Wire fraud if there are some misrepresentations made to the

parents or to authorities? Fourth, professional sanctions, like losing your medical license for attempting something so wild…Fifth, institutional issues. The clinic where this is tried might find its license at risk…Public revulsion or political pressure can bring forth lots of creative sanctions."

These are excellent points, but still it is notable that none of these American laws or potential legal risks is specific for human genetic modification. Further, if the team making a designer baby were to hide their failures and only produce to the world an apparently healthy designer baby, then none of these laws might in practical terms come into play.

It is possible that an American law could be passed in coming years. For example, the US House Subcommittee on Research and Technology recently held a hearing called, "The Science and Ethics of Genetically Engineered Human DNA."[xxv] In addition, a new US House bill is pending at the time of the writing of this book that would prohibit viable human embryo modification.[xxvi] It would also require the FDA to consult religious leaders on evaluating whether to approve three-person IVF in the US.

At the US congressional meeting,[xxvii] CRISPR pioneer Dr. Jennifer Doudna gave testimony along with Dr. Victor Dzau who is the President of the Institute of Medicine (IOM) of the US National Academy of Sciences (NAS), Dr. Elizabeth McNally who is Professor at Northwestern University, and Dr. Jeffrey Kahn, a Professor and bioethicist at Johns Hopkins Hospital. Note that the US NAS will be hosting a meeting on human germline modification from December 1–3, 2015.

You can read Doudna's prepared statement here[xxviii] and watch a video of the entire meeting that I just posted on YouTube.[xxix] All four people who were on the panel were very eloquent and tackled some tough questions.

[xxv] http://1.usa.gov/1G0Xwnr
[xxvi] http://www.ipscell.com/2015/06/congressembryo/
[xxvii] http://www.ipscell.com/2015/06/doudnacongress/
[xxviii] http://1.usa.gov/1QGKmX4
[xxix] http://bit.ly/1FrDvVN

Doudna outlined three central roles for the US to play as CRISPR and human genome engineering technology advance; (1) provide expert information about risks and benefits; (2) lead an international consortium to draft guidelines; and (3) to educate the public about the benefits and risks of genome editing. I strongly support these roles.

Kahn and others discussed ethical issues as well. For example, Kahn said in response to a question about ethics that human germline editing is a line that has never been crossed before. A couple of participants also said that some day, "it may be considered unethical not to use this technology to edit the human germline." If that comes to pass, we could be in a new reality where making GMO sapiens is almost mandated by societal pressure or even by law.

The difficult issue of consent to human germline modification was also raised and someone asked, "how do we think about consent of someone who is not yet born?" There is no clear answer to this, although one panelist suggested it might be within the sphere of appropriate parental consent as guardians of their children.

At a more global level going beyond just the few countries mentioned above, human germline genetic modification is (where laws exist) almost uniformly prohibited. More than forty countries prohibit heritable human modification.[xxx] In most of Europe, for instance, creation of designer babies or even three-person IVF children would be banned.[xxxi] There is almost a universal consensus around the world that germline human genetic modification is not ethically acceptable at this time. Even so, it may well proceed anyway via methods such as CRISPR or three-person IVF either inside or outside the UK. What can science tell us about the risks?

Monkeying around with primate eggs and genomes

In the mid-2000s, Professor Shoukhrat Mitalipov (Figure 4.5) was experimenting with the reproductive cells of macaques, a kind of monkey. Using a cloning-like method in his laboratory in Oregon, he first removed

---

[xxx] http://www.geneticsandsociety.org/downloads/7_Reasons.html
[xxxi] http://conventions.coe.int/Treaty/EN/Reports/Html/168.htm

Figure 4.5. Dr. Shoukhrat Mitalipov, a leader in the area of human therapeutic cloning and three-person IVF/mitochondrial transfer technology to prevent mitochondrial disease. Image credit Natalie Behring, reproduced with permission.

the nuclei of some monkey eggs and then popped into them the nuclei of eggs from different monkeys that acted as donors.

Could the eggs that were missing their own nuclei make do with the nuclei of eggs from these donor monkeys? Could such hybrid eggs grow into actual animals? Mitalipov proved the answer was "yes" and reported the breakthrough in a 2009 paper [7].

He was able to create living macaques that had three genetic parents. There was a mom monkey, a dad monkey, and an egg donor monkey. Some DNA of each of these three genetic "parents" was present in every cell of the newly made monkeys with the donor providing the mitochondrial DNA.

How are Mitalipov's monkeys doing today?

Do these GM monkeys provide hope for GM human studies such as three-person IVF?

His three-parent macaques so far appear to be healthy, which is an encouraging if preliminary finding.[xxxii,xxxiii] It will be important to continue to monitor the health of these primates throughout their lifespans. Mitalipov's team will also examine their fertility as they mature as well as the health of their offspring. Producing a lot more three-parent monkeys, for instance several dozen, could reveal potential health problems associated with this technology or alternatively prove the safety and efficacy of the method.

The monkey data so far is encouraging, but limited. Even so, today in the UK essentially the same method is now approved to try to make GM people who hopefully would have dodged mitochondrial disease as a result.

The future of three-person IVF

We can expect the global three-person IVF debate to continue as the actual work in the UK begins and the discussion over whether to allow this work in the US goes on. In the UK, the HFEA is working out the kinks of the three-person IVF regulations right now. Since Mitalipov has teamed up with other researchers including Hwang in China, where three-person IVF is apparently allowed, there are likely going to be important developments there as well.

Could there be a research race to be first on three-person IVF?

I weighed into the debate on three-person IVF in the UK on my blog and via an open letter to the UK Parliament.[xxxiv] In addition to the HFEA, others in the UK have influential roles in policies on new medical technologies including the Chief Medical Officer of the UK, Professor Sally Davies, who has been a strong advocate of three-person IVF. Professor Davies referred to the media's report of my safety concerns on three-person IVF as "bunk" in the UK House of Lords.

[xxxii] http://www.nature.com/news/2009/090826/full/news.2009.860.html
[xxxiii] http://www.ncbi.nlm.nih.gov/pmc/articles/PMC2774772/
[xxxiv] http://bit.ly/109pSUP

Lord David Alton, an opponent of three-person IVF, contacted me and informed me of the "bunk" comment by Davies. Lord Alton had apparently asked Professor Davies about my safety concerns cited in the UK press, and the response was "bunk". That is an overly simplistic response to a complex situation.

After some debate, the UK Parliament approved the measure to pave the way for three-person IVF. Even so, serious ethical concerns and considerations about three-person IVF remain, and the Council of Europe opposes the use of this technology. They describe this class of experiments as "incompatible with human dignity and international law."[xxxv]

To this day Grifo defends the past efforts to make three-person IVF babies in the US and China even as the safety and efficacy of this IVF-based technology in humans remains unknown.[xxxvi] Grifo sees the UK as being wise in speeding forward on three-person IVF:

> "At least the Brits get it. They will be ahead of the world. It's too bad it's taken so long. It could have been done 15 years ago. "I put my toe in the water and got shark bite [sic], so I'm done with it. It's too bad because it cramps creativity, it inhibits medical progress."

Now with UK approval for this technology, the experiments to test how well it works will be done primarily via making GM humans rather than in animal models such as the macaques, greatly upping the risks. At the same time, the reward could be very positive in the form of the prevention of horrible mitochondrial diseases. In either case, we need to continue the dialogue on the reality that the babies produced will be GMO sapiens.

References

1. Esbjornson, RL Thomas (1984) *The Manipulation of Life*. Harper & Row.
2. Edwards, RGDJ Sharpe (1971) Social values and research in human embryology. *Nature.* **231**:5298:87–91.

[xxxv] http://bit.ly/1GjwLx2
[xxxvi] http://ind.pn/1GaPeys

3. Farber, SA (2008) U.S. scientists' role in the eugenics movement (1907–1939): a contemporary biologist's perspective. *Zebrafish*. **5**:4:243–5.

4. Barritt, JA, *et al.* (2001) Mitochondria in human offspring derived from ooplasmic transplantation. *Human Reproduction (Oxford, England)*. **16**:3:513–6.

5. Ishii, T (2014) Potential impact of human mitochondrial replacement on global policy regarding germline gene modification. *Reprod Biomed Online*. **29**:2:150–5.

6. Annas, GJ (2010) Resurrection of a stem-cell funding barrier--Dickey-Wicker in court. *N Engl J Med*. **363**:18:1687–9.

7. Tachibana, M, *et al.* (2009) Mitochondrial gene replacement in primate offspring and embryonic stem cells. *Nature*. **461**:7262:367–72.

Chapter 5

Build-a-Baby Better via Genetics

"People want access to their genetic information because it is fascinating and meaningful. A few small changes in your DNA can turn your eyes blue, make you lactose intolerant or put some curl in your hair."

— Anne Wojcicki, Founder of 23andMe[i]

The genesis of commercial human genetic testing

Does everything technologically new and powerful have to start in someone's garage? It feels that way at times.

I wrote earlier about how the first commercial GM whole food plant company began in a garage here in Davis, California in 1981. West of here another revolutionary tech company germinated in a garage in Menlo Park, California in the late 1990s. You might know it. Google.

The Chair of Stanford's Physics Department at the time, Stanley Wojcicki, along with his wife Esther, who is an educator, had three daughters: Susan (now CEO of YouTube, owned by Google), Janet, and Anne. Stanford was buzzing with technological innovation at that time.

Google's founders Larry Page and Sergey Brin met at Stanford in 1995 and started toying around with the idea of a web search tool and company.

---

[i] http://www.huffingtonpost.com/anne-wojcicki/secret-code-dna_b_1187141.html

Figure 5.1. Anne Wojcicki, Founder of 23andMe. Image from Wikimedia credited to TechCrunch.[ii]

Their first effort was strangely enough called BackRub. Google itself came into physical being a short time later in Susan Wojcicki's garage in 1998.

The genesis of Google also led to another important, more personal event. Susan's sister Anne met Brin and there was some chemistry there. Eventually the two were married and for years appeared as high-tech's most impressive power couple. In 2008 they had a son named Benji Wojin (his last name being a mashup of his parents'). Notably, Anne and Sergey reportedly donated Benji's DNA to science.[iii]

[ii] http://en.wikipedia.org/wiki/Susan_Wojcicki
[iii] http://www.celebfamily.com/internet/sergey-brin.html

Anne Wojcicki, pictured in Figure 5.1, went on to found 23andMe, a company blazing the trail on direct-to-consumer genetic and genomic testing. 23andMe early on promised a head-to-toe array of genetic insights. For customers interested in their personal genetics and the role of genetics in their family tree, 23andMe sells an analysis service. The 23andMe service depends on customers literally spitting into a tube and mailing it back to 23andMe. The company isolates the DNA from the mouth cells in the saliva for analysis. I myself have had my genome analyzed by 23andMe and found the experience fascinating. 23andMe is very user friendly and welcoming.

23andMe was named one of *Time Magazine's* best inventions of the year for 2008.[iv] The prototypic female customer shown in an envisioned 23andMe pamphlet by *Time* has a number of predicted traits and health risks. The traits range from the rather mundane (wet earwax, average short-term memory, does not have a sweet tooth, average number of freckles or moles) to the more unusual (face does not flush red when she's tipsy, sons have average chance of being bald, good at learning to avoid mistakes).

The health risk assessments from 23andMe are striking in some cases as well: not resistant to malaria, almost no chance of getting Amyotrophic lateral sclerosis (ALS), not a carrier for cystic fibrosis, increased risk of MS, 12% chance of breast cancer, not resistant to HIV. The statement "Unknown odds of being freaked out by these test results" at the bottom is also an important addition by *Time*.

Wojcicki grew up in an academic household with fascinating dinner conversations. In a HuffPost blog piece, *A Secret Code in All of Us*, she relates the first time she heard about genes at age 6 over dinner one night:

"I first heard about 'genes' when I was six years old. At dinner one night, I heard my mom tell my sister, "It's in your genes…She explained that there was a code inside me. That code plus my environment made me who I am. I was amazed. A secret code! In me! This was far better than any mystery puzzle I could have imagined and I was hooked."

[iv] http://content.time.com/time/magazine/pdf/best_invention_2008.pdf

23andMe has had to overcome numerous challenges, including most recently a regulatory tangle with the FDA over its direct-to-consumer genetics service,[v] which left the business model mostly suspended. Unless you were an existing customer of 23andMe as I was, you would not get all those disease and trait predictions until the FDA and the company resolve some remaining issues. Still, there are signs of things starting to be ironed out. In 2015, the company received FDA approval for the first direct-to-consumer test for a genetic disease (Bloom Syndrome), which is a good indicator of things to come for 23andMe.[vi] Another possible challenge is that the personal genetic data collected by 23andMe could raise problems with privacy.[vii]

Wojcicki is likely to continue to be a powerhouse in the area of commercialization of genetics and genomics. She appears to be a true believer that these technologies will transform our reality for the better:

> "Just as computer technology and the Internet created whole new industries and extraordinary benefits for people that extend into almost every realm of human endeavor from education to transportation to medicine, genetics will undoubtedly benefit people everywhere in ways we can't even imagine but know will surely occur."[viii]

23andMe in addition has at times pursued some more controversial directions that suggest it could play a central role in the evolution of genetics-based interventions in human reproduction.

Genetics dating and mating service: I prefer a child with ...

When we evaluate a potential mate, many things come into play both consciously and probably subconsciously. Factors that can figure into such a decision about choosing a possible mate can include physical

---

[v] http://bit.ly/1bH9nOB

[vi] http://bit.ly/1bH9oSF

[vii] http://bit.ly/1jmM1B4

[viii] http://bit.ly/1u79PO2

features, perceptions of intelligence, kindness, sense of humor, and also intangibles such as just one's gut feelings.

Those who study human mating suggest that, taken together, these factors may coalesce around our general perception of what sort of child that we might have with this person and what kind of parent they would be to that child. It is a fairly error-prone approach to picking a mate and producing children.

What if you could bring science, and more specifically genetics, into play when choosing a mate? Would that be more likely to yield better, more predictable outcomes? And what if you could even get a better sense of the potential genetic makeup of your child, were you to choose one from any number of possible mates, before making a commitment?

23andMe has taken a large step in this direction. Via a novel predictive tool, you and a potential mate, or even just you by yourself screening through potential gamete (sperm or egg) donors, could click a button (see Figure 5.2) and have 23andMe predict what your future baby together might be like in terms of specific DNA and traits. For example, what do you prefer, green or brown eyes?

23andMe has patented the idea, sparking some controversy. The patent[ix] by Wojcicki is entitled "Gamete donor selection based on genetic calculations." This has been referred to colloquially as "Build-a-Baby" in comparison to the teddy bear company, Build-a-Bear, where one can design one's own teddy bear.

The 23andMe patent makes an unusual claim as to how a mate selection process might play out and how their tool for this could be used:

> "…to compare the probabilities of observing the phenotype of interest resulting from different combinations of the genotype of the recipient and the genotypes of the plurality of donors to identify the preferred donor…"

What does this mean in plain English?

They are saying that their tool could assist people to select a potential mate from a group of possible mates based on genetics, the more specific

---

[ix] http://www.freepatentsonline.com/8543339.pdf

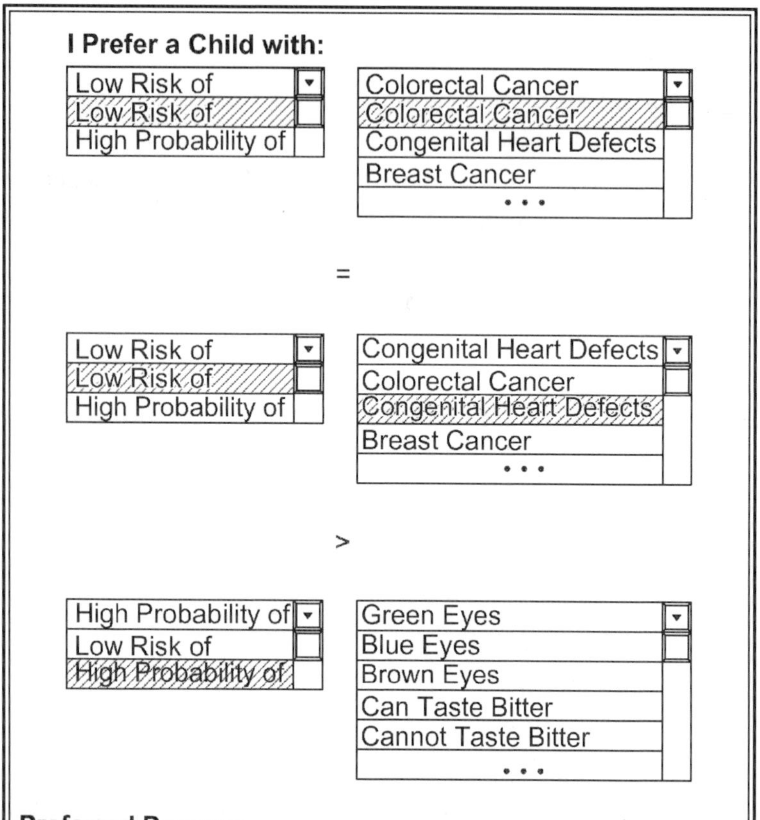

I Prefer a Child with:

Low Risk of ▾		Colorectal Cancer ▾
Low Risk of		Colorectal Cancer
High Probability of		Congenital Heart Defects
		Breast Cancer
		• • •

=

Low Risk of ▾		Congenital Heart Defects ▾
Low Risk of		Colorectal Cancer
High Probability of		Congenital Heart Defects
		Breast Cancer
		• • •

>

High Probability of ▾		Green Eyes ▾
Low Risk of		Blue Eyes
High Probability of		Brown Eyes
		Can Taste Bitter
		Cannot Taste Bitter
		• • •

Preferred Donors:

Donor	Score	Colorectal Cancer	Congenital Heart Defects	Green Eyes
A	80	0.01%	0.005%	65%
B	72	0.015%	0.006%	70%
C	60	0.02%	0.006%	50%

Figure 5.2. Shown is 23andMe's "Build-A-Baby" patent. American patent figure images are considered in the public domain with this one sourced from freepatentsonline.[x]

goal being to select a gamete (meaning sperm or egg), and ultimately a baby, that carries certain "phenotypes" or traits.

For example, if you are a woman wanting to have a blonde, blue-eyed baby boy (a common stereotyped view of what Americans, for example,

---

[x] http://www.freepatentsonline.com/8543339.pdf

might choose to pick) with above average height, intelligence, and health, the tool will point you to the man or men with the sperm with the best chance of having the genetic "right stuff." Genetically-speaking, this donor and you would have the highest probability of together producing such a "better baby," based on your preferences. The donor's genetics are the key. If that sounds to you like at least a shift toward eugenics, then you are not alone. I feel the same way.

In Figure 5.2, you can see a diagram (23andMe calls it "FIG. 4") from the 23andMe patent application in which this type of build-a-baby approach plays out, as described in the original legend written by 23andMe:

> "FIG. 4 is a diagram illustrating an embodiment of a user interface for making user specification and displaying the results. In this example, the recipient has specified that she prefers low risk of colorectal cancer and congenital heart defects equally, and to a lesser degree she also prefers green eye color. A donor selection process such as 200 is performed, and the results page shows preferred donors A, B, and C. For each donor, the statistical distributions of the desired genotypes of the hypothetical child resulting from the combinations of the recipient and the donor, as well as an optional score calculated based on the statistical distribution are displayed. Alternatively, all the donors can be shown in a ranked list."

This sounds about as far away from the romantic nature and randomness of baby generation via normal sex as one might get. On the other hand, it rings a bell if you ever watched that old show the Dating Game in which contestants, if they were female, for example, would choose from Bachelor #1, #2, or #3. The woman had to make this choice without seeing the men and based it upon her blind interview of them, after briefly asking each of them some probing questions.

In the 23andMe's matching, which I call the "Genetic Dating and Mating Game," a woman might choose sperm from Bachelor A, B, or C (the men would not have to be unmarried) without ever seeing these men, but based on the genetic "answers" that 23andMe has derived from analyzing their genomes. The selected sperm that she finds most compelling could then be obtained from a sperm bank for fertilization of her eggs. At one point 23andMe allowed prospective reproductive partners to together link their genetic data profiles on the website to predict traits of offspring, but it is unclear if that option is still available. Still, some people tried out that predictive service to see what sort of kid they might produce together and found it interesting, including Duke Professor Nita Farahany who is an advocate of human genetic modification. Farahany told me in an interview that she and her husband found it fun.

Although 23andMe currently has said it has no plans to pursue the actual use of its gamete selection tool such as at fertility clinics,[xi] there are other players in this area. The GenePeeks[xii] Matchright tool briefly discussed in Chapter 1 has some similarities. California Cryobank, one of the largest sperm banks in the world, also offers customers a variety of tools to select donors. Hence in a sense, their clients can customize their own future offspring made with what they view as the "best" donor sperm based on traits. Going a step further, California Cryobank has a remarkable celebrity look-a-like tool as well.[xiii] Using this tool, customers can pick the sperm of donors who look like their favorite celebrities with the hope of having their child look somewhat like that celebrity.

Changing the genetic equation via designer babies?

In 2014, a company called Cambrian Genomics — previously known primarily for its technology to laser print DNA and also briefly discussed earlier in Chapter 2 — started promoting the idea of using genomics technology to aid in the creation of "perfect" GM babies.

The CEO of Cambrian, Austen Heinz, essentially promoted the goal of "better procreation and living through genomics." Heinz went so far as to say that, in the relatively near future, he could not imagine why anyone will forgo "digitally designing" his or her baby; Cambrian's promotional material even used the key phrase "perfect baby."

It is not a huge leap then to imagine a future in which for-profit companies routinely create GM babies and toward that end they provide

---

[xi] http://bit.ly/1QSgrvg
[xii] https://www.genepeeks.com/
[xiii] https://www.cryobank.com/search?listview=0#

a service to parents in designing the most ideal offspring. While this emerging trend strongly hints that eugenics will play a key role in future human reproduction (see Chapter 7), more immediately it raises many intriguing and disturbing questions.

What is a perfect baby?

Can human beings ever be smart enough to tinker with their own creation without causing disaster? Will we be wise enough to pause for thought and discussion before making GM babies in the first place? I doubt it. Supporters of disability rights also have contended that genetic testing and a quest for human perfection are misguided and could be perceived as part of systemic discrimination against those with disabilities.[xiv] Such so-called ableism could become much more pronounced due to human genetic modification technology.

Some scientists are pushing a new meme as well that has unsettling undertones: democratizing creation.

The idea behind this catchphrase appears to be that anyone should be able to make GM creations. You could think of it as "DIY Creation." Imagine if you could make a tomato plant in your garage that would have rainbow fruit or a potato that would glow so brightly that you could dig it up easily at night from the soil? Intriguing ideas, are they not?

The problem is that creation, even of an organism as simple as a plant, is not something to be monkeyed around with and released haphazardly into nature. There could be serious consequences.

Think that Heinz and Cambrian would defer to the possibility of being slapped on the wrist by regulators? Think again. According to a *Wall Street Journal* piece,[xv] Heinz said: "We want to keep it unregulated...I can't believe that after 10 or 20 years, people will not design their children digitally."

Digital designer babies made by an unregulated process? Some video games allow users to make and customize traits of their characters (or avatars) as well. This can be done with a click of a button from on-screen menus. In the future, will we be able to do the same sort of thing using

---

[xiv] http://bit.ly/1cYEpSN
[xv] http://on.wsj.com/1nLxWgA

menu-based choices for our real-life babies-to-be? It may sound far-fetched, but it feels closer to reality today. For example, while still rare globally, an increasing number of parents-to-be are choosing to select the sex of their future baby.[xvi] In some countries today the level of sex-selection, such as the "son preference" in India and China [1], and the resulting gender imbalances are alarming, as discussed further later in this chapter.

To start to make a GMO designer baby (discussed in more detail in the next chapter), the first thing you would need to do is decide what genetic modifications to make. Changes could range from correcting a disease-related mutation to giving your child certain traits, but to design and make any of these kinds of changes you need information, and in particular genetic and genomic information, as your starting point.

For instance, to design such alterations, scientists need to know where the coding information for those particular traits resides within the genome. While we do not currently have the required genomic and biological information to make very many changes with great confidence, that situation is rapidly changing.

With the plunging costs of genomic sequencing, it is reasonable to expect that by 2020 a million people may have had their genomes sequenced. By crunching that genomic information and correlating individual genetic variants with these people's traits, researchers might be able to come up with a means to make genetic changes that could lead to predictable changes in these traits. On the other hand, such genetic changes could lead to totally unpredictable outcomes. If that risk is too much for you to stomach, such information on correlations between genes and traits could even be used without making gene edits by informing our choices of which embryos to use to make our babies via PGD.

Some genetics and genomics researchers are working feverishly to get better at linking specific genes with traits, such as intelligence. For example, researchers at BGI (formerly known as the Beijing Genomics Institute), the most powerful sequencing company in the world, are on a

---

[xvi] http://slate.me/1PynL9c

quest for what some have called, "the genetics of genius."[xvii] Many researchers over the years have tried and generally failed to link specific genes with intelligence.

However, BGI's Cognitive Genomics group may have a leg (or should I say "brain") up on the competition in the form of a secret "weapon": a collection of DNA samples from hundreds of geniuses. Even so some view this effort by BGI in a very skeptical light, as is evident in a quote from a news story in *Nature*:

> "'If they think they're likely to get much useful data out of this study, they're almost certainly wrong,' says Daniel MacArthur, a geneticist at Massachusetts General Hospital in Boston."

I am skeptical too, but BGI is a very impressive organization so they could make the seemingly impossible or unlikely turn into reality. More broadly, it is not that researchers doubt a genetic link to intelligence, but rather the notion that it can be traced to a few specific genes is viewed as somewhat unrealistic. It is more likely that there are hundreds of genes that work collectively to make the genetic contribution to brainpower. Furthermore, intelligence is also strongly impacted by environment, confounding efforts at predicting or creating intelligence only through genetics.

More broadly, many researchers around the world are very enthusiastic about a new genomics era in which there is more information on our "genotypes," meaning our genetic states and how genetics influence who we are.

In 2014–2015, I interviewed both Church (see Chapter 7) and Heinz (excerpts from the Heinz interview below) on their thoughts about this revolution. Note that Heinz has reportedly recently passed away from suicide and you can read an obituary here.[xviii]

Knoepfler: You have talked about "democratizing creation." What does that mean? What would be the positives of that goal if achieved? Are there risks?

[xvii] http://bit.ly/1bH9qKm

[xviii] http://synbiobeta.com/remembering-austen-heinz/

Heinz: It means that anyone with a smart phone and a credit card can build new living organisms that never previously existed.

The impact will be to allow billions of people around the world to write the code of life. This will have a profound effect on culture, politics, and religion but more importantly it will give a tremendous boost to researchers attempting to solve the world's greatest problems.

The risk is non-existent if the organism does not leave a locked down centralized Bio-safety level 3 or level 4 facility.[xix] We are currently in discussions with contract research organizations that either have or will have this capacity so that we can enable literally anyone in the world anywhere to make real new life forms on their phone or tablet.

It's important to note that even though the poor will now be able to write the code of life just like the rich do now — only the rich will be able to access their creations. That's because the cost to release organisms is extremely high because of the need for significant testing and government interaction to clear regulatory hurdles.

We hope that we can not only democratize creation but also democratize access to creation through our Creature Creator crowdfunding platform.

Knoepfler: What kinds of genetic changes are customers wanting to make in living things or do you imagine them wanting to make?

Heinz: Since we are all running faulty DNA code we all have a strong personal imperative to fix our code. In our lifetime we are likely to get cancer which is the result of DNA mutations...

Knoepfler: One can imagine one kind of change such as correcting a mutation, but what about what some might view as "novelties" such as a glowing pet bird? We can see Glowing Plant doing this sort of thing with plants. What about glowing animals? Researchers have made GFP animals of various kinds for experiments; what about customers doing it just for fun or the thrill of being a creator?

Heinz: Sometimes things that look like toys are actually critical to the progress of the human species. The gaming industry has definitely pushed ahead microprocessor development and virtual reality gaming is now

---

[xix] http://en.m.wikipedia.org/wiki/Biosafety_level

forcing electronics engineers to solve deep problems that will be useful for much more than games.

Knoepfler: You mentioned that you wouldn't want your customers doing "bad" things. Do you have a bioethicist on board to help with that or an external advisory committee or something like that?

Heinz: Simple the CDC has a list of bad things like Small Pox that we don't print. If we don't know what it is we don't print it. Also we only ship to validated university and commercial laboratories.

Knoepfler: You were quoted in the WSJ, "I can't believe that after 10 or 20 years people will not design their children digitally." Are you an advocate for such human genetic modification? Some folks are scared of the idea of "designer babies"? What are you views on this?

Heinz: My view is that if you don't want a baby with a disease like Harlequin-ichthyosis — I urge you to look it up[xx] — then you will use a computer and genomics to help make your child.

Already people are designing babies digitally via selection whether through pre-implant genetic diagnosis or through mitochondrial embryo transplant i.e. three parent babies.

Knoepfler: Are genetically modified humans definitely on the horizon? Not a question so much of "if," but rather "when"?

Heinz: Genetically modified humans are already in existence in many countries. In fact thanks to a more relaxed medical regulatory environment in Europe you are about to see large numbers of genetically modified humans there. Glybera, from Dutch biotech firm UniQure is the first approved gene therapy treatment in the Western world…

It's likely that in the next 100 years almost every human will be genetically modified throughout their lifetime. But already you have to wonder how having these new genetically modified humans will begin affecting culture and politics.

My cofounder George Church at Harvard University pointed out this irony recently. Imagine the leader of a well established anti-GMO group in Europe gets the treatment. Then you'd have a genetically modified

---

[xx] http://en.m.wikipedia.org/wiki/Harlequin-type_ichthyosis

organism rallying against the existence of genetically modified organisms. It's going to be fun to watch culture and politics change in real time in light of these absurd cases.

Knoepfler: How do you counter bioethicists such as Marcy Darnovsky who calls your vision "techno-libertarianism" and raises ethical concerns about genetic modification?

Heinz: I think she has a good point. If the power of synthetic biology was released without any regulation I can't imagine things going well.

Despite this last answer, Heinz did not appear particularly enthusiastic about his industry being regulated very much.

Preimplantation genetic diagnosis (PGD)

If we think about the apparently most legitimate use of potential human genetic modification being the correction of disease-causing genetic mutations, it is important to discuss a powerful alternative to gene editing: preimplantation genetic diagnosis (PGD).

What is PGD and how does it work?

During the PGD process, couples that are facing potential genetic disease in their offspring (or who choose to do it for other reasons that might be more ethically complicated) undergo IVF to produce embryos in the laboratory. These embryos develop, going from one to two, to four to eight cells, in a dish.

When an embryo has divided into eight cells, one or two of its cells can be gently removed to use for PGD (see Figure 5.3). This is also called "blastomere biopsy" because at this stage each cell is called a blastomere. Blastomeres can be used for PGD or to make human ESC lines [2]. In principle, this is done without harming the embryo. PGD can also be done later at the blastocyst stage when there are more cells.

Regardless of the stage at which cells are harvested from embryos for PGD, DNA is isolated from embryonic cells for analysis. Genetic mutations can be identified by polymerase chain reaction (PCR), a method that amplifies specific sequences of DNA, or by genomic sequencing. PCR and DNA sequencing are so sensitive now that the process is usually successful even if started with just one or two cells. The remaining six-to-seven cell

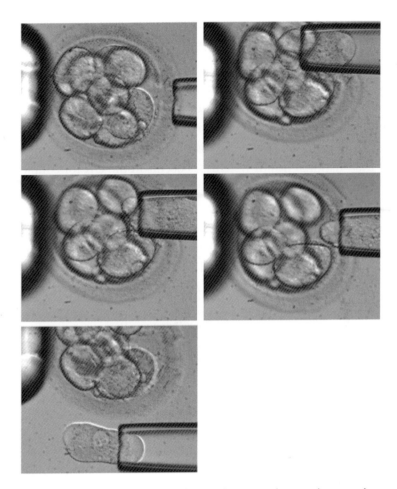

Figure 5.3. Sequential photos of an early stage human embryo undergoing the removal of a single cell (from left to right and top to bottom). The removed cell can be used to make embryonic stem cells or for preimplantation genetic diagnosis (PGD), where its genome is analyzed for mutations. Image credit: Dr. Robert Lanza. Reproduced with permission [2].

embryo can then sometimes, but not always, continue to develop normally. The loss of the one or two cells is a stress that some embryos at this stage can compensate for, leading ultimately to the birth of a normal baby.

For example, if a couple has produced ten embryos and five have a mutation, as determined by PGD, reproductive efforts could be focused on the subset that are "normal," meaning those that lack the mutation, and that were not irreparably harmed by losing the few cells needed for PGD. A PGD procedure, sometimes also referred to as Preimplantation Genetic Screening or PGS, costs approximately $5,000-$7,500 in the US. When you factor in that it can often take three or more attempts to achieve a successful pregnancy through IVF, and then factor in the potential added PGD costs at each round, it all could add up to a very expensive proposition. However, it is provided free in some countries as part of their national health systems, as occurs in the UK for couples with certain genetic conditions.

There are hundreds of disease-causing mutations that could be excluded in this way, including those for Huntington's disease, sickle cell anemia, hemophilia, cystic fibrosis, mitochondrial disease; the list goes on and on when you also include genetic mutations that predispose people to certain diseases, such as the *BRCA1* breast-cancer associated mutations, and more.

If one were attempting to make GMO sapiens via CRISPR-Cas9, PGD would also likely need to be used anyway to screen embryos for those that both have had a successful gene-editing event and do not have other genetic errors introduced into them by the gene-editing tools. In such cases, not only would specific target genes (e.g. the one mutated in cystic fibrosis) need to be analyzed, but also the entire genome would need to be sequenced to look for gene-editing mistakes potentially introduced elsewhere (more on this risk and screening in the next chapter).

In 2000, Jacques Cohen along with another assisted reproductive technology pioneer, Dr. Santiago Munne, first founded a preimplantation genetics company called Reprogenetics.[xxi] Now PGD is done at hundreds of locations around the world.

CRISPR versus PGD

There has been much concern expressed in 2015 about how gene-editing technologies, such as CRISPR, might be used prematurely in the clinic in

---

[xxi] http://reprogenetics.com/

an unsafe or unethical manner in humans to try to prevent genetic diseases or enhance traits. However, in a way the dialogue on this usually misses a crucial, more basic question.

Why would anyone even try human gene editing in the germline given the existence of the very powerful, already proven safe technology of PGD?

In a match up of say CRISPR versus PGD, PGD would win almost 100% of the time as by far the best choice to tackle genetic disease. It would be both safer and more effective. Right here in Davis and Sacramento, for example, the local fertility clinic called California IVF[xxii] offers PGD for a huge number of genetic diseases.

As I am writing this now in 2015, the biomedical sciences community has been abuzz with the publication of two papers related to the use of gene-editing technology to prevent human genetic diseases. There was the paper from the team in China on the use of CRISPR to genetically modify human embryos for a hypothetical future path to treat beta thalassemia ([3] mentioned earlier in the book and see more in the next chapter). Then there was a paper from the Salk Institute on the use of gene editing TALENs to try to prevent human mitochondrial disease [4]. The authors in both cases talked about potential future clinical applications of gene editing, but realistically why go that route instead of PGD? Most of the time there would not be a compelling reason to do so.

What are the challenges or shortcomings of PGD? PGD does not always work, and some biopsied embryos fail to develop due to the removal of their cells. There are also ethical concerns over the use of PGD for sex selection or potentially in ways that are eugenic.

Still, it is a generally safe and effective technology. It works for the diseases that were focused upon in these two gene-editing papers: beta thalassemia and many mitochondrial diseases.

There are only a few hypothetical, very rare scenarios where trying gene editing instead of PGD would possibly make sense. Below I discuss the three that most readily come to mind.

[xxii] http://www.californiaivf.com/genetic-diagnosis-PGD-CGH.htm

A parent is homozygous for a dominant mutation. We all have two copies of each gene. If both copies of a given gene are of the same type in our genome, we are called "homozygous," while if the two differ in some way, we are called "heterozygous."

Sometimes genetic diseases can manifest when only one of the two copies of a gene is mutated, that is, even if the other copy is still normal. This is called a dominant mutation. Other genetic diseases will manifest only if both gene copies are mutated. This is called a recessive mutation. If one parent is homozygous for a mutation, meaning they have two mutant gene copies, then 100% of their embryos and children will inherit one mutant copy of that gene (assuming their reproductive partner does not have the same mutation). If, in addition, the mutation in question is dominant then 100% of that parent's offspring will inherit the disease. In this extremely rare sort of situation, there are no "normal" embryos to find by PGD. Gene editing could hypothetically have a unique positive role in this sort of case.

Both parents have a mutation in the same disease-causing gene. Another possibility where PGD would not be particularly effective is if both parents-to-be are carriers of mutations. In this scenario, PGD would only very rarely be effective at identifying an embryo that lacks any mutation, as about three out of four embryos would have at least one copy of a parental mutation. Still, PGD might work if enough embryos can be generated. Keep in mind also that for human gene editing, such as that achieved using CRISPR, a large number of embryos are likely to be needed as well as not all will have the correct genetic modification. Some might also receive the edit in the wrong place (an off-target effect) or end up being chimeras where only some cells of an embryo have the gene edit.

One or both parents have multiple kinds of disease-causing mutations. If parents-to-be have multiple genetic disease-associated mutations, then the odds of successful PGD become longer. In such scenarios, in theory at least, gene editing could be used to try to correct multiple mutations in one embryo. For example, in cells in the laboratory and in model organisms, CRISPR has been proven in some cases to successfully edit multiple genes at once. At the same time, attempting to gene edit more than one mutation could raise possible risks such as for off-target effects where one mistakenly edits the wrong gene.

In the end, even in a hypothetical future scenario with an essentially perfectly accurate gene-editing technology, opting for PGD is going to be the wiser choice for parents and doctors almost every time. The reality of PGD as a competing and generally superior technology to human genetic modification needs further discussion. One of the top review articles on the potential therapeutic use of gene editing never even mentioned PGD once [5]. Including PGD in the dialogue could help to temper imprudent considerations of rapidly employing CRISPR, or other gene-editing technologies, for clinical use in humans.

Sex selection

One of the more ethically complicated applications of PGD technology today is the selection of embryos based on their sex. PGD can very readily determine the sex of embryos and allow couples to have a baby with essentially a 100% certainty of it being either male or female. Sex selection has a legitimate use in PGD specifically when parents-to-be face the prospect of a sex-linked genetic disorder. In these genetic diseases, depending on whether a mutation sits on the X or Y sex chromosome, children of a certain sex will always have the disease, so picking a child of the other sex via PGD is an important and effective way to avoid the disease. However, there are other uses of sex selection that are more controversial [6].

Some couples may seek to have a more "balanced" sex distribution in their families and so use PGD for that purpose to direct future births to be a boy or a girl. Later in the reproductive process, certain couples may combine ultrasound with abortion to try to achieve sex selection as well.

In addition to the PGD family balancing act, some have suggested parents should be allowed to use sex selection for other, personal reasons because of a principle of "reproductive liberty" [7]. Could this notion of reproductive freedom come to include human genetic modification? It is possible.

Other parents-to-be may seek a child of a certain sex for cultural reasons, and in certain cultures male or female children are preferred.

Some countries by law do not permit the non-medical use of PGD for sex selection. For instance, the sex selection of embryos by PGD is illegal in the UK.

While I personally do not believe that a human embryo is a human being, nonetheless a human embryo is something special. Therefore, trashing embryos in biohazard waste just because they are the "wrong" sex is unethical unless there is a compelling health reason to select one sex over the other.

Sex selection either by genetics or other practices also poses serious societal problems. In her fascinating and provocative book *Unnatural Selection: Choosing Boys over Girls and the Consequences of a World Full of Men*, Mara Hvistendahl reports an in-depth analysis of the causes and consequences of sex selection [8]. She quotes the gut-wrenching estimated statistic of 163 million females missing from Asia due to different forms of sex selection. It remains unclear how the countries most affected by this staggering gender imbalance, including China, India, and increasingly other South Asian nations, will cope.

Savior siblings

Some families finding themselves in the position of having a family member with a severe illness, such as a blood cancer, may choose to produce another child and use PGD specifically for so-called "HLA matching." The goal is to have a new child that is perfectly immunologically compatible to the existing child or other family member who is ill, so that a transplant can take place. Rather than wait for the rare chance of finding a potential future, unrelated bone marrow or stem cell donor who is matched to the sick family member, the family intentionally makes an HLA-matched child via IVF and PGD. They do not just want to have a child. Rather, the intention is to create and use the future child for a medical purpose. These children are often colloquially referred to as, "Savior Siblings" [9].

Should parents have the right to create a child so that it will serve as a medical donor to a sibling? If there is no HLA match through such efforts, would it be permissible to create an HLA matched donor sibling via genetic modification technology?

The savior sibling topic was tackled directly in the book/movie, *My Sister's Keeper*. In the film the parents of the character Anna Fitzgerald create her as a savior sibling for her sister Kate. She is expected by the family to donate a kidney to Kate, and was created for that purpose via IVF. Therefore, she is a genetic match to Kate, who is gravely ill.

This scenario creates tension and sets up the key question of whether such a person created in this way would be under an obligation to act as an organ donor. How about the engineered person's feelings about this and about her or his own creation? This type of situation also raises the dilemma of how GMO sapiens would react to the knowledge of their unique status. Could they also be discriminated against or be created for a specific purpose?

Human GMO economics

An important question is whether GMO sapiens could become part of a corporate enterprise intended to generate profits. Although genes themselves at this time are not patentable in the US, due to a 2013 Supreme Court decision,[xxiii] GMOs can be patented and frequently are. Humans, whether "wildtype" or GM, cannot be patented.

Another more likely possibility is that the specific technologies, devices, and processes used for the different steps involved in producing a GM human could be patented. Indeed there are already patent applications pending or approved for mitochondrial transfer/three-person IVF technology and for CRISPR-Cas9 technology used for gene editing (also discussed in Chapter 2). For the CRISPR-Cas9 inventions, multiple competing applications exist and ownership of the intellectual property related to CRISPR-Cas9 remains uncertain at this time.[xxiv] The future commercialization of potential tools for gene editing in humans are wrapped up in possible financial and corporate interests.

[xxiii] http://nyti.ms/1zTJXcz
[xxiv] http://bit.ly/1FvEibb

If businesses come to own the technologies used to produce GM humans, it is possible that at some point they may exert a great deal of control over how the technology is used and could challenge attempts to regulate its use for specific applications. If GM human technology takes off it could become a multi-billion dollar industry, perhaps often affiliated with assisted reproduction clinics. In his book *The Science of Human Perfection*, Nathaniel Comfort recommends caution for consumers considering use of cutting edge technologies [10]:

"As biomedicine and healthcare become increasingly important in daily life, a healthy skepticism becomes literally vital. It can help us benefit from the powerful new knowledge biomedicine daily produces. As patients and consumers we must that knowledge intelligently — lest other interests trump our own."

If you think about the corporate, profit-driven priorities of GMO giants Monsanto and Syngenta, which incidentally could end up merging into one even bigger GMO colossus,[xxv] it is not too difficult to imagine a GMO sapiens industry with great power. It is even possible that one of these two plant GMO companies could take an interest in human modification.

There could be other businesspeople who jump on this bandwagon as well. The entrepreneurs and computer techies of Silicon Valley love to design and build things. Imagine designing and building a new life form? They are already excited about this idea. Google, in its new incarnation as part of parent company Alphabet is now reportedly interested in genetic modification of organisms. For technophiles, making a new type of human could be the ultimate challenge and may provide a fix for their desire to leave their mark on the world. It is not surprising then that synthetic biology and human genetic modification have captured the attention and in some cases the big money of folks in Silicon Valley.[xxvi] Transhumanism is also very popular there.

[xxv] http://on.wsj.com/1QS7UIF
[xxvi] http://slate.me/1zbRUW3

More broadly, there are a growing number of people interested in the commercialization of human genetic modification technology. This goes well beyond those already mentioned to companies that might be more inclined to directly attempt to edit human germ cells, embryos, and people.

Let us take a look at a couple examples of ventures that combine the idea of helping people via biotechnology, including potentially via gene editing, with that of making big bucks too.

OvaScience

OvaScience[xxvii] is ostensibly a company built just on the idea of treating infertility. If you go to their website the front pages splashes, "A WOMAN'S BIOLOGY IS EXTRAORDINARY" in all caps with the subtitle, "New Fertility Treatments for Women." Behind these words is an image of a very healthy-looking, attractive woman. So far, this is not so different from the webpages of hundreds of other fertility clinics. However, OvaScience has embraced new technology in ways that other clinics have not. OvaScience is working on precursor or stem cells that can give rise to human eggs. They call these special, powerful cells "EggPC cells," which they rightly indicate could be the basis of new approaches to treating female infertility. At the same time, EggPC cells could be used to produce human eggs for research and for other applications, including human genetic modification.

OvaScience could potentially make a lot of eggs from each customer, perhaps hundreds or thousands of eggs from just one woman via her EggPC cells. With the ubiquitous nature of sperm, it could be a fairly straightforward matter to then produce hundreds or thousands of human embryos per customer. We could think of these as "candidate" embryos. While OvaScience's technology has not yet produced entirely functional human eggs, it appears like that this will happen soon.

OvaScience has "seen the light" on the power of this technology — it goes well beyond simply treating infertility. Part of their business model looks to be an intimate connection between fertility and human

---

[xxvii] http://www.ovascience.com/

enhancement. Antonio Regalado wrote of this connection in a 2015 piece in Technology Review:[xxviii]

> "OvaScience has been collecting, and studying, what it believes are egg stem cells from the outer layer of women's ovaries. The company has not yet perfected its stem-cell technology — it has not reported that the eggs it grows in the lab are viable — but Sinclair predicted that functional eggs were 'a when, and not an if.' Once the technology works, he said, infertile women will be able to produce hundreds of eggs, and maybe hundreds of embryos. Using DNA sequencing to analyze their genes, they could pick among them for the healthiest ones."

That sure sounds like the genetic dystopian film *GATTACA* becoming a reality. *GATTACA* is discussed in more depth in Chapter 8.

How would the "healthiest" be defined? How would this technology be regulated to ensure appropriate ethical, medical and scientific oversight? The "Sinclair" referred to in the quote is anti-aging specialist and Harvard scientist David Sinclair, who also appears to be a proponent of human modification according to Regalado. Sinclair is on the Advisory Board of OvaScience along with Dr. Grifo, who did the controversial oocyte manipulation experiments discussed in Chapter 2.[xxix] Sinclair goes on to invoke the idea of designer babies as quoted in the same article:

> "Genetically improved children may also be possible. Sinclair told the investors that he was trying to alter the DNA of these egg stem cells using gene editing, work he later told me he was doing with Church's lab … His goal, and that of OvaScience, is to 'correct those mutations before we generate your child,' he said. 'It's still experimental, but there is no reason to expect it won't be possible in coming years.'"

Correcting mutations "before we generate your child" is another way of saying making a GM person. The idea of reverting mutations is frankly more seductive than PGD. In a hypothetical future with GMO sapiens,

---

[xxviii] http://bit.ly/1aQ7QpI
[xxix] http://www.ovascience.com/about-us/advisors

companies such as OvaScience would be focused on making a profit from modifying people, which raises its own set of ethical issues. OvaScience has formed a joint venture with another business, genetic modification company Intrexon, to specifically use gene-editing technology to prevent human disease, and to make investors very happy by generating big profits.

Mitogenome therapeutics

In addition to patenting the three-person IVF technology, Shoukhrat Mitalipov has created a for-profit biotech called Mitogenome Therapeutics to commercialize his three-person IVF/mitochondrial transfer technology. In this version of three-person IVF, instead of nuclear transfer, Mitalipov's team usually does what is called, "spindle transfer," where just the DNA itself wrapped in a spindle (a structure used by cells to move their DNA around) is transferred into the donor oocyte that has had its own nucleus removed. After cell division, a new nucleus can form with the transferred DNA inside.

A challenge for Mitogenome is that the FDA currently prohibits the use of this technology in humans in the US. However, perhaps in part as a means to get around this regulatory limitation, Mitalipov and Mitogenome have recently teamed up with the Chinese mega-biotech firm, BoyaLife, to do various work together, including three-person IVF.[xxx] Such work is apparently thought to be permissible in China.[xxxi] It is worthwhile to consider whether there might be a competition between Mitalipov in China and UK researchers to try to be the first to successfully conduct three-person IVF (for the moment not including the earlier ooplasm transfer work of the infertility researchers in the US in the 1990s).

The cooperative joint venture between Mitogenome and BoyaLife also involves Korean cloning researcher, Hwang Woo-Suk, well known for his controversial and fraudulent research on human therapeutic

---

[xxx] http://www.ipscell.com/2015/02/mitalipov-hwang/
[xxxi] http://bit.ly/1MdiIw0

cloning, which led to the retraction of the high profile publications reporting his "findings" and to an international scandal discussed in Chapter 3. More recently, Hwang has gained attention by working with a team to bring the Woolly Mammoth back to life via de-extinction. *Science Magazine* reported on the collaboration involving both Mitalipov and Hwang in this way:[xxxii]

> "The newspaper says initially work will focus on animal cloning but eventually move on to work with human materials. Mitalipov's 'strength is in primate stem cells. My specialty is in cell nuclear transplantation. So we've agreed that if we combine his strength with mine, we can create a breakthrough outcome in curing maternal line genetic disease, on which he is now focusing,' the paper quotes Hwang. Hwang said they will place their laboratory in China to avoid Korea's strict bioethics regulations."

Avoiding US and Korean ethics regulations may be good for corporate profits, but what about for the people involved and for the future GM humans to be created? It remains unclear why Hwang is part of the Mitogenome-BoyaLife collaboration, but reportedly Hwang remains interested in human cloning, raising the potential worry that at some point an effort might be engaged to clone actual human beings.

Is money a motivating factor for Mitalipov and his home institution of OHSU? There are mixed signals on this question:[xxxiii] However, having the intellectual property rights for this technology and for human therapeutic cloning methods could prove lucrative in the future if these approaches find additional applications in medicine and become more widespread. Especially if genetic modification services get folded into fertility clinic offerings, substantial profit may be generated.

In terms of predicting a price tag for the production of future GM humans (e.g. even for a "simple" mutant gene correction), the best

---

[xxxii] http://bit.ly/1MdiIw0
[xxxiii] http://www.wweek.com/portland/article-23100-splice_of_life.html

high-tech, genetics-based medicine to look at as an example is the first gene therapy medicine, Glybera,[xxxiv] cited in the quote by Heinz earlier. This treatment is designed to combat a genetic disease that affects how people metabolize lipids. The price per treatment is the most for any medicine in history: $1.5 million USD. I would expect that germline human genetic modification by CRISPR would be at a minimum in the same ballpark and could easily run into millions of dollars.

Genetics and human choices

Overall, genetic and genomic technologies today provide the public with an unprecedented amount of information about their own DNA and even that of potential mates, other family members, and possibly also that of their future children made with different candidate reproductive partners. It is very exciting, but also unsettling.

When combined with advancing assisted reproduction technologies, the new world of powerful genetic and genomic innovation has brought us to the brink of a *GATTACA*-like situation where it is at least plausible that a non-sexual means of having children could in the future become a preferred and relatively much more common choice for parents-to-be.

In today's culture in which more information, and getting that information more quickly, are viewed as a high priority, it is easy to see how what we view today as an "artificial" means of reproduction — human genetic modification — could rather rapidly come to be viewed as the norm with profound implications for humanity. That has happened over the years with IVF and PGD, which are now widely (although not universally) viewed as acceptable and within the norm.

Genetic tourism

There is likely to be a demand for human germline therapies both for the prevention of genetic diseases and for human enhancement. There may

---

[xxxiv] http://reut.rs/1HG7Qoo

be social pressures on parents to go ahead and try to make GM children. These kinds of forces could include competition from social group members who are also parents.

In some countries, there is no specific law to prevent the production of designer babies, so that is not a definite roadblock. That door is open. This technology combined with different regional regulatory and legal perspectives establish the groundwork for a new phenomenon that I call "genetic tourism."

In this potential scenario of genetic tourism, people would travel from one place to another in order to purchase genetic modification procedures. Genetic tourism could involve *GATTACA*-like modified forms of IVF in which PGD is used to select only specific embryos that meet certain non-medical, narrow genetic criteria. In that sense, one type of genetic tourism may already be happening now with sex selection by PGD being currently permitted in certain countries such as the US.

Some fertility clinics today may already be going much further by using PGD to sell "better" embryos and future children to parents-to-be in certain countries around the world. The broader trends bundled under the umbrella term "medical tourism," where patients travel to countries that have much less regulatory oversight, are already causing some serious problems.

Much the same as medical tourism, genetic tourism more specifically could raise ethical and legal concerns. Genetic tourism customers may find themselves in situations of legal limbo should something go wrong. This is particularly the case should an adverse outcome occur down the road, years after a procedure, as genetic tourism purchases would be unlikely to include long-term follow-up. The biggest risk to clinics willing to participate in genetic tourism might be bad publicity, leading to reactive regulatory action after the fact. For example, if after a GM baby is born with defects or if parents are unhappy with the results and seek legal redress, this may attract the attention of regulatory authorities.

Another type of medical tourism is reproductive tourism, which bears some similarities to predicted genetic tourism. In reproductive tourism, customers travel to countries with either weaker regulatory oversight or cheaper prices for reproductive procedures. Reproductive tourism has a number of associated negative outcomes including commodification of women involved in surrogacy or the human egg market. Reproductive tourism is a growing, multibillion-dollar industry that is focused on providing customers with babies and certain kinds of "better" babies [5]. It is not difficult to foresee a future multibillion-dollar genetic tourism industry that in effect sells a turbo-charged version of the same services and "products" in the form of GMO sapiens babies.

Are designer babies the next step on the GM timeline?

I have made an overall timeline of roughly the last 40 years of events in the areas of stem cell research, cloning, gene therapy and genetic engineering, giving preference to the key events that have brought us to the brink of creating GMO sapiens (Figure 5.4). Hopefully, you can get a sense from this of just how fast these technologies are progressing and how there is crosstalk between them that is enabling of possible future production of GM humans.

We have seen in the evolution of genetic engineering a progression from cutting and pasting DNA molecules in tubes in the laboratory to genetic tinkering in bacteria, to changes in individual mammalian cells in the laboratory, to the creation of GM plants, and to the production of GM animals. At the same time DNA sequencing, stem cell, cloning, and reproductive technologies have sped forward as well.

The table is set, so to speak, for the next step to be creation of GM humans in coming years. The tools and know-how are there. I would imagine the demand will be there too, ranging from therapeutic goals (for example, to correct genetic disease) to enhancement such as trying to make a smarter or better-behaved kid.

In the next chapter I discuss how one might go about trying to making a GMO sapiens. I explain the exciting technology involved in straightforward terms as well as detailing the potential benefits and great risks involved.

Chapter 6

DIY Guide to Creating GMO Sapiens

"What I cannot create, I do not understand."

— Richard Feynman, Caltech Professor

How would one go about creating a GMO sapiens?

Would it be difficult? What steps would be involved? What level of understanding would be needed?

Could you do it DIY-style, maybe even in your garage with some equipment purchased on eBay?

It would be a challenge, but with say a hundred thousand dollars, basic cell and molecular biology laboratory equipment, and a partnership with a fertility clinic, it would be doable. This chapter is not intended to enable people to literally try to make a GMO sapiens, but rather to illustrate that the technology already exists and that it could be attempted by any of the hundreds of laboratories (probably not garages) around the world. Whether for the correction of genetic diseases or for the creation of designer babies for enhancement, the steps involved to give it a try are fairly straightforward even if doing it in a responsible manner could be far more difficult or even impossible.

As discussed in other chapters, there are few laws or binding regulations that would prevent it from happening here in the US and in numerous other countries.

Why make a GMO sapiens and what is your target?

The first thing to do when contemplating such a project would be to ask yourself why you want to do it.

For fun? For fame and glory? The novelty of making a new human?

Since we are talking about playing around with creation and making a customized human being, I do not believe those reasons are good enough.

Even the goal of making a "better" human being is highly questionable, given the risks involved and the potential personal and societal outcomes that might be awaiting such a creation.

If we turn to literature — and, yes, fiction is fiction — there is still a sort of collective, powerful wisdom there that "messing" with the stuff of life and playing God is risky business, even in the real, non-fictional world. See more on the perspectives of culture and literature on human modification in Chapter 8.

If despite cautionary notes, you still wanted to try to create a modified human, how would you go about doing it? Where would you start?

The first scientific step is to figure out in a more specific sense what you want to achieve. You would address this question in a preliminary design phase. Just as an example, let us say that you want to create a very specific genomic change that is associated with a strongly reduced risk of a negative health outcome: susceptibility to HIV, malaria, Ebola, breast or prostate cancer, a lung disease, or a neurological disorder.

First, you pick a condition to fix. Then, you have to find out if there is a specific place in the genome that is convincingly associated with it that you could edit. If so, then you need to learn as much as you can about that region of the genome. Time for some homework. Taking on more risk, it is also possible that instead of focusing on a disease, your goal is to enhance traits, such as intelligence, appearance, and so on. In that case, your homework would be more difficult, as you will have to try to determine all of the genetic elements that associate most strongly with those traits. There could be dozens or even hundreds.

If you are focusing on disease prevention, a mutation in a specific gene often causes diseases, and you therefore would have a genetic bull's eye to aim for. There are a number of public databases where you can get much of the information that you would need on the gene in question, including publications about it in the biomedical literature database called PubMed.[i]

You can learn about the protein made by the gene there as well. A good depth of understanding is important to be able to do this in the most careful manner possible. You might learn on PubMed how the mutation in the gene you are interested in changes the resulting protein.

For example, hypothetically focusing on cystic fibrosis and the *CFTR* gene that is mutated to cause this disease, you could search the scientific article database PubMed for "CFTR gene" to get all the information that is currently available.[ii] Such a search yields more than 4,000 results. You do not have to look at all of those, but digging into some of them will teach you a lot. You can get the genomic sequence of the gene at any number of websites, including through the NIH gene search tool,[iii] again with the example of *CFTR*.

Since you are going to be focusing on human beings, you will want to include the word "human" in your search or look for "*Homo sapiens*" in your results. You can learn some of what you need to know from a genomic and genetic angle about CFTR here,[iv] including the location and DNA sequence of the place in which it is mutated in cystic fibrosis. However, some reasonable knowledge of genetics and biology would be required to make sense of the information you obtained from your searches. Another catch is that each person's vesion of any given gene will vary in sequence by about 0.1%.

The CRISPR connection

Once you have done your homework and thought out the genetic target to change in your future GM baby, next comes the production process. However, some more advanced design work is needed here as well. There

---

[i] http://www.ncbi.nlm.nih.gov/pubmed

[ii] http://www.ncbi.nlm.nih.gov/pubmed/?term=cftr+gene

[iii] http://www.ncbi.nlm.nih.gov/gene/?term=cftr+human

[iv] http://www.ncbi.nlm.nih.gov/gene/1080

are a number of methods out there to make genetic modifications, but the CRISPR-Cas9 approach so far appears to be the simplest, most affordable, and easiest to put together in terms of the specific tools needed. CRISPR-Cas9, cited by *Nature* magazine in its list of top science innovations in 2013,[v] is a simple, but powerful new laboratory technology for editing a genome. A key part of using CRISPR is to design the targeting part that acts as a GPS to ensure the gene edit occurs in the right place in a vast genome.

Where did CRISPR technology originate? The earlier research that led to this technology was not focused on gene editing at all. This is often the way of things in science; as you study one thing, you somehow discover another. Rarely, as with CRISPR, this thing is of huge significance for something else entirely. CRISPR came from studies of the bacterial genome.

Microbiologists doing research on *E. coli* and other bacteria, starting in the 1980s, turned up these odd repeat sequences in the bacterial genome [1]. Genomic repeats are a common occurrence in a wide range of organisms, but these were odd in that the repeated clusters had a regular spacing and were very short. What could these repeats mean? Were they some kind of code?

They remained entirely a mystery even a dozen years later [2] and still when they took on the CRISPR name (short for Clustered Regularly Interspaced Short Palindromic Repeats) a few years after that. Multiple different teams including one working on, of all things, cheese and yogurt (which are made using bacterial cultures), started uncovering the function of CRISPR and an associated family of factors called Cas — the most famous member of which is Cas9.

The yogurt researchers realized that the CRISPR system could be used in a clever way to solve a pesky problem they faced.[vi] Viruses not only plague us, but also infect bacteria as well, including the specific beneficial bacterial strains that are essential for making yogurt and cheese.

---

[v] http://www.nature.com/news/365-days-nature-s-10-1.14367
[vi] http://bit.ly/14zL7uD

These viral infections of bacteria, therefore, are a serious concern for yogurt and cheese makers.

Two scientists, Philippe Horvath and Rodolphe Barrangou, were doing research in this area. They were working at the company Danisco that, amongst other things, makes cheese and yogurt. Horvath and Barrangou figured out that CRISPR was used by bacteria as a way of immunizing themselves across generations against viral infections [3]. As discussed earlier in this book, CRISPR repeat sequences act as viral fingerprints that allow the bacteria to mount an immune response to viral infections. The CRISPR-Cas9 system effectively chops up the viral genome, thus protecting bacteria from viral infection.

In this way, bacteria use CRISPR-Cas9 as a natural anti-viral defense immunity mechanism [4]. The bacteria fight off viruses just as we humans have ways to battle viruses or might be immunized to not get sick from them.

It may come as a big surprise to you as it did to me that many of the bacteria used to produce an array of dairy products today are not naturally occurring. Researchers at DuPont, which bought Danisco, specially bred bacteria over the past eight years to have extra CRISPR sequences in them. To achieve this, the scientists exposed bacterial strains to viruses and selected the ones that had the greatest resistance due to extra CRISPR sequences. As a result DuPont's special bacteria that are now often used to make yogurt have immune systems super-charged against viruses. Ironically, this is all thanks to CRISPR research even though the bacteria in question were not directly genetically modified.

CRISPR as a laboratory tool

It turned out that CRISPR-Cas9 could be used in other brilliant ways beyond the world of dairy products. Clever researchers figured out that man-made, adapted versions of CRISPR-Cas9 tools could be produced to work in mouse and human cells as powerful gene-editing tools. The CRISPR part is the guidance system that takes the machinery to the desired place in the genome, where the Cas9 nuclease enzyme cuts the DNA in that spot. During the fixing of the cut spot and at that location in the genome, one can intentionally introduce specific genetic modifications.

Figure 6.1. Professor Feng Zhang, CRISPR-Cas9 innovator, pictured in his laboratory. Image is copyright Kent Dayton, used with his permission.

Massachusetts Institute of Technology (MIT) Professor Feng Zhang (Figure 6.1) has published key work in this area [5]. He has filed a patent and notably had it awarded[vii] via an accelerated patent review process. Professors Jennifer Doudna and Emmanuelle Charpentier of the University of California Berkeley (UCB) and University of Vienna, respectively, also came up with the idea of using this technology in different applications [4]. They have also filed patents based on this technology. Another CRISPR researcher, Virginijus Siksnys, deserves some cedit too.[viii]

CRISPR-Cas9, as adapted for use in the laboratory as a gene-editing tool, is as mentioned earlier in the book similar to a smart Swiss Army

[vii] https://www.google.com/patents/US8697359
[viii] http://wrd.cm/1NerhJO

Figure 6.2. The author's conception of how CRISPR-Cas9 can be compared to a personified Swiss Army Knife superhero editing the genome.

knife for DNA, with a magnifying glass (genomic scanner), a scissors (to make cuts in DNA), and a pencil to write in new DNA units or base pairs (see Figure 6.2). This tool knows just where to go in the genome, cuts there, and then, "writes" new DNA code.

What this means is that via this technology, biomedical scientists can quickly and easily edit the genomes of cells, such as ESCs, and in turn even entire embryos or organisms, such as mice. Or People. Other new adaptations of CRISPR-based technology are already coming out that can do more than just delete genes or make edits. For example, there are new

CRISPR methods for turning the transcription of specific genes on or off instead of editing them.

The CRISPR-Cas9 technology is being commercialized by a number of for-profit ventures, including Caribou Biosciences, CRISPR Therapeutics, and Editas Medicine. These CRISPR companies will compete with existing zinc-finger nuclease technology that is used to do the same kinds of things. *Nature* quotes a leader in the zinc-finger arena as follows:

"'CRISPR has taken the academic world by storm, and it's a very exciting new technology,' says Philip Gregory, chief scientific officer of Sangamo. But, he says, there are still several kinks to be resolved, including studies that suggest that Cas9 can make cuts at off-target sites in the genome. Zhang says that his group and others are working on increasing the enzyme's specificity, and have already made some gains."

Gregory shows CRISPR some respect, but also rightly points out some hurdles remain for this technology. The same applies to Sangamo's own type of gene-editing technology, which is based on non-CRISPR-based DNA-binding tools including so-called zinc fingers. Like everything in life, there are challenges to getting CRISPR to work precisely for any given application. Still most experts believe the edge goes to CRISPR over other gene-editing technologies.

What would one do to get the CRISPR-Cas9 tools needed for human genetic modification? One place to turn is the repository called Addgene,[ix] which has all the necessary CRISPR-Cas9 plasmids.[x] Plasmids are circular segments of DNA that are molecular and cellular manufacturing powerhouses. In the case of CRISPR-Cas9, you need plasmids to make the Cas9 nuclease (the scissors that cut the genome) and to make other components of this system, including the so-called "guide RNAs" that are the scanner-like component that take Cas9 to the right place in the genome (and hopefully to only that chromosomal address).

[ix] https://www.addgene.org/

[x] https://www.addgene.org/crispr/

Addgene has the plasmids needed and will ship them at a relatively low cost (e.g. $60 each). Although Addgene has a verification step for new accounts and one must list an institutional affiliation, it is not clear to what extent this might be an obstacle to someone wanting to make GM people, who might already have an institutional affiliation or simply invent one.

Addgene also provides fantastic advice and guidance on how to design and conduct one's CRISPR-Cas9 experiment.[xi] In the example of the mutant *CFTR* gene, you would want to set up a CRISPR-Cas9 system to revert that mutation back to the normal or "wildtype" sequence. You would need to design this system so as to have the smallest possible chance of it making genomic edits in the wrong place. Such an incorrect edit in the wrong place or a faulty change at the right place in either case could cause a new disease.

A key step in the process of making your CRISPR-Cas9 system is the design of the so-called guide RNAs that take the machinery to the correct genomic address. Fortunately, there is a simple, yet powerful tool to aid you in this process, hosted by MIT.[xii] To my knowledge, there are no restrictions on who can use the MIT online design tool.

Could someone really try to use CRISPR in humans clinically or for enhancement? There is a sense that such controversial experiments could begin in the near future. For example, in their book *Evolving Ourselves* [6], authors Juan Enriquez and Steve Gullans discuss possible clinical uses of CRISPR and sound convinced that this sort of use is imminent:

> "In-vivo human trials are certainly not too far off. CRISPR is not squir-relled away inside top-end secret labs; it's in the hands of high school and college kids; it has permeated scientific research…"

I have to agree with them on this. In addition and more importantly, thousands of skilled scientists in laboratories around the world also have this technology. In my own laboratory, just about everyone is

---

[xi] https://www.addgene.org/crispr/planning-your-experiment/
[xii] http://crispr.mit.edu/

doing something with CRISPR and often in human cells. Some of these laboratories may choose to pursue heritable human genetic modification experiments.

The Addgene collection of CRISPR plasmids, all in some way related to CRISPR-Cas9 technology, includes more than 500 different types of plasmids.[xiii] Remarkably in 2014 alone there were, according to Addgene's own statistics,[xiv] more than 17,000 requests for CRISPR-related plasmids. That is a great deal of gene editing going on around the world, and in the laboratory context, it is almost entirely unregulated. George Church has said that a CRISPR laboratory could be set up for as little as $2,000.[xv]

In the laboratory, begin at the beginning

To make a GM organism of any kind, you need to start early in the organism's development process. To produce a GMO sapiens, you could start with the germ cells (the egg or sperm, although an egg could work best since the sperm genome is so compactly packaged that it would be very difficult to edit it directly). Or you could begin with a single-cell embryo made by IVF. Another option is to go back further still to edit in the so-called primordial germ cells (PGCs) that eventually become sperm and eggs, although human PGCs and their differentiation constitute a relatively new area of research that is still being worked out.

Why edit so early?

Starting at the germ cell or at the one-cell embryo stage would currently be the best way to get all the cells of the resulting human body genetically modified in the same way. Otherwise, for example, if you started by editing an eight-cell embryo, you might end up with a human mosaic or chimera, in which some cells have one genome while others have a different version of the genome (with or without the gene edit). While we all likely have a few cells in our bodies that bear slightly

xiii http://bit.ly/1QSZOfk
xiv http://bit.ly/1QSZOfk
xv http://bit.ly/1cjL4qp

different DNA sequences, and thus humans are "microchimeras" in a sense, the dramatically more extensive mosaicism that would be created by a faulty gene editing process could well lead to disease. Having many cells with different genomes in a single person could make her or him sick due to a variety of problems, such as immune issues.

A la carte genetic modifications

While most would only consider gene-editing technology for its therapeutic potential, at some point it will likely go beyond that to attempts at human enhancement. For instance, knowing human nature, some parents, given the opportunity, are likely to consider genetically modifying a baby for non-medical reasons.

GMO sapiens could be produced with the goal being that they are more attractive or taller and have more muscle mass. In the latter case, such super-sized GM humans could well be at increased risk of other diseases, such as cancer. We would not know until someone experimentally created them and they could be studied over time.

Another possible kind of enhancement attempt could come from some parents-to-be wanting smarter babies. It is debatable whether scientists trying to make such GM geniuses could readily and reproducibly achieve that goal. There would be numerous risks as well, including accidentally producing children with impaired cognition or altered behavior. You might well produce a less intelligent child by mistake or one that has autism or prematurely develops Alzheimer's.

One might imagine more complex negative outcomes too where the primary goal of increased intelligence is achieved, but always results in antisocial or violent behavior: think a super-villain from a comic book or movie (discussed more in Chapter 8). Furthermore, even if the goal of increased intelligence is successfully achieved and there was no other significant negative new attribute, would the boost in raw brainpower come with a balancing effect of increased wisdom? Perhaps not.

It is also likely that some parents would ask for other genetic modifications that are highly questionable, including specific eye, hair, or skin colors. It would not shock me if some parents-to-be decided that they wanted glow-in-the-dark children. In earlier chapters, I discussed the whole spectrum of fluorescent animals out there, ranging from GM mice to monkeys that contain an amazing jellyfish protein that glows in the dark under certain conditions, called green fluorescent protein (or GFP).

Could someone make a GFP person just for the heck of it? There are also other fluorescent proteins available, allowing GM animals to be made that glow in other colors as well. These proteins include yellow, red, and blue just to name three out of a whole rainbow of possibilities for fluorescent GMO sapiens. Think human equivalents to the GloFish mentioned earlier. Such efforts would be wildly irresponsible.

One might imagine that perceptions of ideal qualitative or quantitative traits would vary between different cultures. In Western cultures, there could be an expectation of certain harmful but perhaps popular "Hollywood" or "Ken and Barbie" type of desired outcomes. Some may not necessarily just want taller children, but also children who would grow up with different sized specific body parts as adults.

Larger muscles, breasts or penises. Smaller stomachs (reduced abdominal body fat). Longer legs. Permanent "bee-stung" lips seen on supermodels. Other options may be available, including altering specific facial features, such as creating certain kinds of cheekbones, chins, hairlines, and such. Additional selections might be offered such as eliminating male pattern baldness, reducing body hair, and lowering the numbers of moles. This last one could even be considered a medical selection given the link between moles and skin cancer.

If you think these ideas sound far-fetched, consider that Americans alone spend tens of billions of dollars each year on plastic surgery procedures and creams to try to achieve these kinds of goals. Some of the time elective cosmetic surgery is done on children. In the future, we might have "cosmetic genetic surgeons" who do "surgery" on our family's genes for cosmetic reasons. In other countries the sensibilities and cultural expectations could lead to other kinds of genetic modifications of humans for "enhancements."

A potential role for stem cells

Human genetic modification might be achievable using stem cells and adaptations to cloning technology. Today, two specific types of powerful stem cells can be made from just about any person: induced pluripotent stem cells (IPSCs) and embryonic stem cells (ESCs). IPSCs can be made via a process called "cellular reprogramming," while, in theory, ESCs can be produced via therapeutic cloning from any person.

Both IPSCs and ESCs are pluripotent, meaning that they can form any type of human cell in the body except for the extraembryonic tissues, like the placenta. Totipotent stem cells can make both an entire embryo and extraembryonic tissues. The extraembryonic tissues are not directly needed to make an embryo itself, but are required for an embryo to develop into an actual person during pregnancy. Therefore, in principle it might be possible to use IPSCs or ESCs made from a specific person to either clone that person or, with an intervening CRISPR-Cas9 gene-editing step, to produce a genetically modified clone-like version of that person. However, to make that happen such cells would first have to be injected into an existing embryo to generate the extraembryonic tissues.

Why possibly use stem cells instead of germ cells or embryos for human modification? Even in healthy women, human eggs are hard to come by and egg procurement is an ethically complicated process. Some women have few if any healthy eggs as well. Stem cells might be a work-around in such cases, and IPSCs and ESCs could in theory be produced in a limitless supply. While sperm are highly abundant in most men, healthy sperm can be in short supply for others. Sperm are also not amenable themselves to genetic modification as mentioned earlier due to the particularly compact nature of their DNA. Therefore, using stem cells might be a relatively efficient way to produce GM babies for infertile couples.

While in most cases the use of stem cells to produce a designer baby would entail an intermediate step — the creation of sperm or eggs from pluripotent stem cells (or from the PGCs discussed earlier) — it is also possible that with advances in stem cell technology researchers could, in time, consistently produce totipotent stem cells that are genetically matched to adults. In this scenario, such cells might then be genetically modified before developing into both an embryo and its extraembryonic tissues (including the placenta and umbilical cord). This approach could in theory allow GM human offspring to be produced. In making GM mice for research, scientists always start the process with some kind of stem cells.

Build a better mouse and they will come: the experience of making a GMO

What is it like to create a GMO? What thought and technology goes into it and how does it feel when one is faced with one's own GM creation? As a scientist, I have made different kinds of GM mice over the years and studied them. Other scientists in my laboratory are also studying GM mice. We do this to advance knowledge of mammalian development and disease.

As a postdoctoral fellow, a step in science akin to an apprenticeship before becoming a professor, I led an effort in Bob Eisenman's laboratory that successfully made GM mice that lacked a specific gene, called N-Myc, just in the stem cells of the brain, which are called neural stem cells [7]. When we make GMOs in the laboratory lacking a specific gene we call them "knockouts." What would be the effect of the knockout of N-Myc?

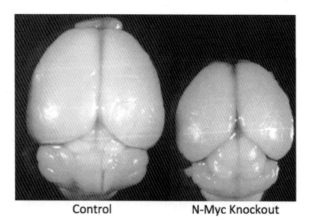

Control   N-Myc Knockout

Figure 6.3. Brains isolated from normal sibling and N-Myc knockout GM adult mice, shown at the same magnification. Image credit: Paul Knoepfler.

You can see from the pictures of the N-Myc knockout mouse brains that I took and that are shown in Figure 6.3 just how small they are compared to normal mouse brains taken from the unaffected siblings of the knockout mice.

Although all the other genes in these GM mice were genetically normal, the mice exhibited profound changes that were strikingly similar to a human condition called microcephaly. People who have microcephaly possess an unusually small head and brain. While some of those with microcephaly have normal intelligence, most have some level of cognitive impairment.

The causes of microcephaly in people were not well understood. It turned out that our N-Myc knockout mice were collectively a useful tool for understanding the causes of human microcephaly. They also provided insight as to why people with too much N-Myc often get abnormal brain overgrowth in the form of devastating brain tumors.

Only a few years after I published my paper on the N-Myc knockout mice with microcephaly, an article was published by a different group studying human genetics in which they described mutations in the human N-Myc gene (akin to our mouse knockout) that occurred in people with a developmental disorder called Feingold Syndrome [8]. The human Feingold Syndrome patients who had mutant N-Myc all had microcephaly just like my N-Myc knockout mice. This was a revelation for me as to the power of knockout mice for understanding human development and disease. Score one for GM animals in science.

Overall, my studies of the N-Myc GM mice provided important insights into normal brain development and also into that of brain tumors. As a scientist, that outcome felt, and still feels, like an important accomplishment. At the same time, on a personal level I admittedly felt a bit unsettled for a time by my initial foray into creating a GMO in the form of these mice. You alter or remove a gene and you can change a creature's brain size? That was positive scientifically and at the same time uncomfortable on some levels personally.

Thinking about making GM humans raises the stakes and risks in terms of possible unintended consequences. The makers of designer babies could well feel more than unsettled by their creation of GM human "children," particularly if something goes wrong.

What if you make a mistake?

If you make a mistake when creating a GM mouse, the consequences could be serious. You may have wasted a lot of time, money, and effort by altering the wrong gene or modifying your gene of interest in the wrong way. You have also done animal research without a clear benefit. In principle, such a mistake could cost you your career, depending on where you are at in your development as a scientist. Another type of "negative finding" would be that you may have done the gene editing perfectly, but even so nothing happens in the knockout organisms.

Making a mistake in producing a GMO sapiens could lead to far more disastrous outcomes. Think that GMO mistakes are unlikely? Scientists are too smart to make such mistakes? Think again.

Geneticist Craig Venter is, in my opinion, one of the edgiest biological innovators of our time. Widely hailed as unique — part scientific trailblazer, part entrepreneur — Venter pushes the boundaries of what is possible. For example, he led the team that created the first synthetic bacterial chromosome.

Venter's team is about as good as they come at doing experimental science, but even they are only human. They made a surprising mistake when tinkering with genomics to make the first synthetic genome. Recall the Feynman quote at the beginning of this chapter: "What I cannot create, I do not understand." Feynman has many famous quotes, but this one is near the top. It was scrawled on his blackboard at Caltech.

In a tribute to Feynman, Venter's team wanted to put in a new code of sorts for Feynman's quote inside the synthetic genome that they made, except they goofed up and Caltech noticed. The bacterial DNA-encoded message from Venter's team read by mistake: "What I cannot build, I cannot understand."[xvi] This was a subtle, but significant difference from Caltech's perspective. A goof.

If even Craig Venter can make such a mistake, genomic and reproductive scientists seeking to make designer babies can make mistakes too and even if they do not, the tools they use to make genomic edits,

---

[xvi] http://bit.ly/1H9ZZRM

such as the CRISPR-Cas9 machinery, can foul up sometimes. A mistake in genetic modification as small as the Venter team error could have huge consequences for health in a GM person.

Getting back to Feynman, I would argue that we cannot really understand human genetic modification until we create GM humans, and then if things go wrong, it may well be too late in terms of long-term negative consequences. So there may be a paradox, as only in the building and creation of designer babies can we come to understand the outcomes and what might go wrong, as well as whether it is wise to even try to make a GMO sapiens in the first place.

What could go wrong

Since we are talking about literally cutting and forcing changes in the human genome, things easily could go wrong. I can foresee three main types of errors occurring in the actual gene-editing process (Figure 6.4), along with other possible problematic outcomes, which I discuss later in this chapter.

In what I call a "Type 1" error, the gene-editing tool has gone to the right place in the human embryo's genome, but has made the wrong edit once there. This sort of error could partially disrupt, inactivate, or change the normal function of the target gene in question, leading to a new disease. I view this type of error as being like

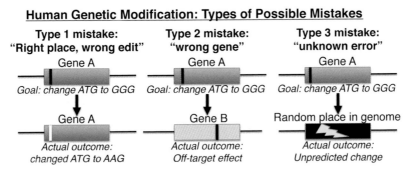

Figure 6.4.   Some of the possible kinds of errors that human gene editing could cause. Image created by the author.

a typo — one with potentially disastrous consequences. Yes, such an error instead may have no functional consequence, but do we really want to take the risk?

In a scenario with a Type 2 error, the gene-editing tool has gone to the wrong place in the genome. We scientists call these Type 2 errors "off-target" mistakes. For example, your goal might be to change a mutation in the *BRCA1* breast cancer-associated gene, but instead the gene-editing tool has made a change to an entirely different gene. This could lead to trouble and may also be difficult to detect.

Online gene-editing predictive tools are available that researchers can use to plan their genetic modification experiments. These tools can sometimes predict possible off-target sites that might be likely to be the site of a gene-editing mistake. But, in principle, such a "wrong location" mistake could happen anywhere in the genome and at unpredictable sites. Therefore, when modifying a human genome in a clinical setting, it would be prudent to sequence the entire genome to look for such mistakes, as well as to check whether the desired edit itself did in fact occur. Recall again that the human genome has billions of units in it, so looking for mistakes is a tough task.

Another issue arises as well. When we are talking about targeting a single cell (e.g. the fertilized egg) for a genetic change, it is not entirely clear how one could go about getting the whole genome sequence data without destroying the embryo. One would likely have to wait until the embryo had grown to at least the eight-cell stage before removing one or two cells for use in whole-genome sequencing validation experiments in a PGD-like procedure. As a result, we would not know our starting point in terms of the genome.

A further challenge is that if the gene-editing tool made a small Type 2 mistake, such as by introducing a single nucleotide edit in the wrong gene or even in an important gene regulatory region that nonetheless is not inside of a gene itself, it might fly under the radar. Can the resolution of even today's robust whole-genome sequencing technology confidently find such a needle in the haystack of a tiny mutation in an entire genome and distinguish it from either simple sequencing error or human genetic divergence? This technology is so new that the best answer at this point to such a question is, "we do not know."

In the third type of mistake, Type 3, things are even less predictable. The gene-editing tool in this scenario has gone to some random place in the genome and made potentially a large mistake, such as deleting an entire gene or even a multigene region. Another Type 3 error would be a potentially even more catastrophic mistake: creating what is called a chromosomal translocation, which occurs when a piece of one chromosome is cut and pasted onto another different chromosome. Such errors may in theory be easier to find because of their relatively big size, but they also pose potentially bigger risks for human health.

The old saying goes that "to a hammer, everything looks like a nail." For CRISPR-Cas9, despite its unprecedented precision compared to past technologies, the same may hold true as for that hammer. For the Swiss army knife that is CRISPR-Cas9, the genome is sure full of places that look like they would be good to cut. I imagine if you did an experiment yourself of walking around your home for an hour carrying an open pair of scissors that at times you would be tempted to cut certain objects. Try it.

Another possible bad outcome could occur even if the gene-editing tools do their job perfectly. In this class of errors, the gene edit that we think will lead to one sort of functional outcome instead leads to another very different one. This is an error of outcome prediction. For example, if we are correcting a mutation in the cystic fibrosis gene, that same "correction," even if successfully achieved, could lead to some unintended change in the function of a nearby gene or to some other DNA element that produces an unhealthy outcome. This could happen if, as is sometimes the case, the On-Off switch for a different gene resides inside of the gene that we are trying to target. Such overlapping genomic machinery is not uncommon.

In his commentary piece entitled *Brave New Genome* in the *New England Journal of Medicine* [9], Professor Eric Lander raises a number of concerns about human genetic modification including unintended consequences:

"Some observers might propose reshaping the human gene pool by endowing all children with many naturally occurring 'protective' vari-ants. However, genetic variants that decrease risk for some diseases can increase risk for others. (For example, the *CCR5* mutations that protect against HIV also elevate the risk for West Nile virus, and multiple genes have variants with opposing effects on risk for type 1 diabetes and Crohn's disease.)"

Additional kinds of human error could come into the picture even if all the gene-editing tools work flawlessly. With all the embryonic manipulation and handling that would be involved, particularly if human genetic modification becomes popular and clinics are making many GMO sapiens babies at once, parents could end up getting the wrong baby. Someone else's baby. Or the parents could get only have "half" the right baby if, for example, the wrong sperm sample was used. As it turns out, IVF mistakes do happen on a regular basis.[xvii] It is also possible even if they get the right baby, it could have had the wrong gene edit introduced into it. Maybe your neighbor wanted a green-eyed baby and you wanted a baby with long legs, and the laboratory accidentally made the wrong changes in the wrong embryos. Consider, for example, that sometimes doctors will accidentally cut out the wrong kidney and you can appreciate that such "simple" avoidable mistakes do happen.

Could we correct a mistake?

What could we do if an error in genetic modification happens or if an unexpected negative result occurs during the process of trying to make a GMO sapiens?

Generally, if we detected a mistake in gene editing of the kinds discussed above, what we might do about it would largely depend on the nature of the error and when it was detected.

If the mistake was discovered in a preimplantation embryo that was at an initial blastocyst stage or even earlier, then I expect that the team doing this work would simply stop the process and most likely either discard the embryo that has the error or study it to determine how the error occurred and with what consequences.

[xvii] http://dailym.ai/1DZzsPL

If an error was detected following data analysis after an embryo has been implanted into the mother, then things are going to be far more complicated than if a problem surfaced earlier in an embryo in a dish in the laboratory. The pregnancy could be terminated if it was early enough, but that itself raises ethical questions.

It is also possible that genetic modification mistakes might not be detected until it is "too late," such as late in pregnancy or even after the birth of a GMO sapiens baby. At that point after birth, the hope would have to be that the new person with the laboratory-introduced genetic mistake would not suffer ill health due to the error. This also would raise profound ethical questions.

After the fact, it would probably be impossible to correct such a mistake. People just have too many trillions of cells. Therefore, if the error led to severe health consequences or even to death, society and the media would view it as a catastrophe. In addition to harming the GMO sapiens involved and one might say their family as well, this turn of events could jeopardize this entire field of research including that limited to the laboratory.

What would happen if (or should I say "when"?) a designer baby is made with a mistake or two or three in its genome? What would be the biological consequences?

We cannot be entirely sure until the designer babies are made and the mistakes inevitably crop up. Therefore, a la Feynman to know how this would play out, we would have to pull the trigger and give it a try. I do not think we should pull that trigger, but others are more enthusiastic about doing so.

If mistakes do cause problems, the most likely type would be developmental disorders. Somehow, the glitch in the DNA introduced by the scientists by mistake could cause anywhere from a developmental hiccup to a full-blown failure of human development.

Major developmental problems are often the cause of failed early pregnancies and miscarriages that result. During normal, naturally occurring pregnancies such outcomes might not attract much attention. In the case of a woman implanted with a designer baby embryo, however, the scientists and doctors involved would be carefully watching with ultrasound. They probably would know if and when something major went wrong. If the pregnancy failed, the embryo or fetus would be retrieved for analysis. Would the world ever hear about it? It is hard to say. If the team doing this and the individuals involved told the world that they were making such an attempt, then the world would be watching.

Finally, it is reasonable to think that some biological manifestations of gene-editing mistakes in designer babies might not appear until these GMO sapiens had grown up and possibly had children of their own. Problems might not show up until the GMO sapiens start aging. Things like increased risk of cancer or rapid aging are not out of the question with gene edits, and these diseases could take so long to manifest that we do not learn from them in time to inform subsequent decisions about making more GM humans.

In short, this all points to the fact that making designer babies would be above all else a high-risk, transgenerational experiment. Anyone who has spent time in a laboratory knows that most experiments either do not succeed or yield results that were unexpected. On top of it all, the person at greatest risk would be the GMO sapiens and that person cannot consent because, at the beginning of the experiment, they do not exist. Their children and all future generations in that family would also be affected (barring some kind of genetic reversal editing attempt via assisted reproduction) and would not have consented. It is unclear how this dilemma concerning consent could ever be resolved.

Practical challenges to making a GMO sapiens

Several major logistical and practical problems face those who might want to make GMO sapiens. I have outlined the top five challenges in Figure 6.5.

1. Unknown genomic sequence of the embryo to start. The process of making GMO sapiens would likely begin with a fertilized egg (a one-cell embryo). Unfortunately, as mentioned earlier you cannot sequence its genome without destroying it, but without sequencing it you do not know your starting point sequence. Without knowing the starting

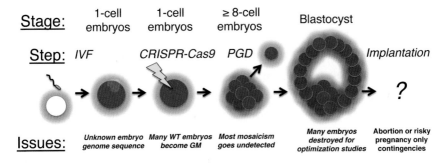

Figure 6.5. Author's illustration of the key obstacles to responsibly using CRISPR to make GMO sapiens.

genomic sequence, how can you design your CRISPR system and how can you be sure later as to whether or not you have created gene-editing mistakes? You cannot. By analogy, this situation of not knowing the starting genomic sequence of the embryo is akin to having to edit a book, but without being allowed to read it first. You are only given a copy of a very similar book or perhaps a different edition. We do have something called the "reference" human genome sequence (a baseline sequence of the human genome from only one or a few individuals) and humans have remarkably similar genomic sequences overall, but we could not confidently predict the precise sequence of the embryo in advance.

2. Mutating normal embryos. In most clinical scenarios where gene editing might be used, one parent has a disease-causing mutation and one does not. This means that after IVF, 50% of the embryos created would have the mutation and 50% would not and would be normal. Therefore, you are inevitably going to put CRISPR into some normal human embryos and may mutate them as well. These new mutations could even cause diseases. Is that ethical? Again keep in mind that you are probably going to have to implant some of these embryos into a surrogate mom without being totally sure of their starting or edited genomic sequence in advance.

3. Mosaicism goes undetected. Upon reaching the eight-cell stage, or later, you would need to do PGD to determine the embryo's genomic sequence and whether the specific genomic modification had worked. To this end, you remove one or two cells and sequence their individual genomes to try to monitor how well your CRISPR has worked. Since you do not screen the remaining cells in this embryo (because you would destroy it by doing so), you may well miss mosaicism where only some of the cells in the embryo were edited appropriately or had off-target effects.

4. Destroying large numbers of human embryos. To optimize gene-editing tools for clinical use in human embryos intended to produce babies, you are likely to need to "practice" on thousands of embryos to perfect the methodology. These tests may well be done with no intent to use these embryos to produce babies. Is that ethical? And where do you get all those human eggs and embryos?

5. Trapped choosing the lesser evil post-implantation? If a problem arises once a GM human embryo has been implanted into a surrogate mother, the options for fixing that problem are limited. Do you continue a high-risk pregnancy, for example, with a fetus that has a genome that was improperly edited as an embryo and now has issues, or do you abort? It appears like an ethical trap as both possibilities could be viewed as unacceptable. If such problems are relatively frequent, is it OK to essentially cause many abortions as part of this clinical experiment?

Overall, heritable human genetic modification is not to be taken lightly. There is enormous power in this technology, but at least for the foreseeable future there are also equally huge risks and associated very complicated ethical questions.

In the next chapter I discuss two movements, eugenics and transhumanism, which in their own ways have embraced the idea of making better humans.

References

1. Ishino, Y, *et al.* (1987) Nucleotide sequence of the iap gene, responsible for alkaline phosphatase isozyme conversion in Escherichia coli, and identification of the gene product. *J Bacteriol.* **169**:12:5429–33.

2. Mojica, FJ, *et al.* (2000) Biological significance of a family of regularly spaced repeats in the genomes of Archaea, Bacteria and mitochondria. *Mol Microbiol.* **36**:1:244–6.

3. Barrangou, R, *et al.* (2007) CRISPR provides acquired resistance against viruses in prokaryotes. *Science.* **315**:5819:1709–12.

4. Jinek, M, *et al.* (2012) A programmable dual-RNA-guided DNA endonuclease in adaptive bacterial immunity. *Science.* **337**:6096:816–21.

5. Cong, L, *et al.* (2013) Multiplex genome engineering using CRISPR/Cas systems. *Science.* **339**:6121:819–23.

6. Enriquez, JS Gullans *Evolving Ourselves: How Unnatural Selection and Nonrandom Mutation are Changing Life on Earth.*

7. Knoepfler, PS, PF Cheng, RN Eisenman (2002) N-myc is essential during neurogenesis for the rapid expansion of progenitor cell populations and the inhibition of neuronal differentiation. *Genes Dev.* **16**:20:2699–712.

8. van Bokhoven, H, *et al.* (2005) MYCN haploinsufficiency is associated with reduced brain size and intestinal atresias in Feingold syndrome. *Nat Genet.* **37**:5:465–7.

9. Lander, ES (2015) Brave new gsenome. *N Engl J Med.*

Chapter 7

Eugenics and Transhumanism

"To know the worst as well as the best in heredity; to preserve and select the best — these are the most essential forces in the future evolution of human society."

— Henry Fairfield Osborn, Eugenicist, *Science Magazine*, 1921.

"The bold code of the transhumanist will rise. That's an inevitable, undeniable fact... We are the future. Like it or not. And it needs to be molded, guided, and handled correctly by the strength and wisdom of transhuman scientists..."

— Zoltan Istvan, Transhumanist candidate for US President, in *The Transhumanist Wager*, 2013

Eugenics takes root in California

Philanthropist Charles M. Goethe was a nature lover who lived in Sacramento where my research laboratory is located. Today's environmentalists, especially those here in Northern California, might be spellbound by the idea of Goethe on first glance. He was a conservationist, who loved the soil and trees, and was interested in education as well. He even founded Sacramento State University.

In short, he had big bucks and was not afraid to part with some of them to help nature and foster education. When you dig just a little more, though, you easily unearth a deeply troubling side to the man. In addition

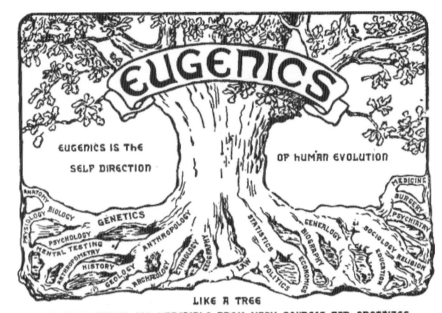

Figure 7.1. A eugenics symbol represented in the Second International Eugenics Congress Logo from the Second International Exhibition of Eugenics, which was held in the fall of 1921, in the American Museum of Natural History, New York. Image source: Wikimedia Commons.

to advancing environmental goals and education, Goethe also had some disturbing ideas and used his power to try to make them a reality too.

Goethe was one of the top eugenicists in America and the world, in the first half of the 20th century. He believed in the eugenics-based idea that there were good and bad people and that the bad ones should not be allowed to freely "breed." Ironically, given Goethe's soft spot for trees, a common eugenic symbol was that of a strong tree with roots in different disciplines of science (Figure 7.1). The idea conveyed by this symbol was "self-direction of human evolution" towards a better future via eugenics based on hard science.

Eugenic notions that most of us now can only read about with abhorrence were surprisingly commonplace back then amongst Anglo-Saxon culture in America and Europe [1, 2]. Even so, today a movement called transhumanism, particularly popular with some in the high-tech crowd in Silicon Valley, embraces some of the same ideas as eugenics.

California was a leader in eugenics or, to be more specific, in a form of eugenics that we might now view as "negative eugenics." This form of eugenics was founded again on the principle that people with "worse" characteristics should be discouraged or even forcibly prevented by law from having children. This would, in the eugenicists' view, make the human race "better." How do you prevent certain people from having children? Forced sterilization was the method of choice, but in the future I imagine the possibility of genetics-based reproductive quarantine also coming into play. This could especially be the case if we make GMO sapiens, some of whom could be considered a danger to society if they have what are perceived to be abnormal, heritable, and negative traits.

Goethe was a fan of forced sterilization and was a leading American advocate of this extreme form of human social engineering well before the Nazis. However, he later felt the German "model" had become even better and should be one that America might emulate. According to reporter Chrisanne Beckner[i]:

> "Goethe mentioned that 'Germany in a few months outdistanced California's sterilization world record of a quarter-century... We must study Germany's methods.'"

The idea of forced sterilization was to enhance the overall human genetic stock by stopping reproduction of "inferior" people. Eugenicists embraced the idea of "better" people breeding more as well. How one back then defined "inferior" and "better" was largely non-scientific and would have been hugely influenced by one's own upbringing and environment.

According to Edwin Black in his book, *War Against the Weak*, Goethe continued to admire the German eugenics efforts and their broader policies, even as the violence perpetrated by the Nazis escalated:

> "Again and again," Goethe wrote to a Nazi eugenicist in early 1938, "I am telling our people here, who are only too often poisoned by anti-German

---

[i] http://bit.ly/1Pyo301

propaganda, of the marvelous progress you and your German associates are making." And once violence broke out in 1938 against Jews in what is now known as *Kristallnacht*, Goethe wrote again: "I regret that my fellow countrymen are so blinded by propaganda..." [3]

It was not just Goethe either. Many leaders in California in the first half of the 20th century supported the eugenics movement, including reportedly leading members of the California Board of Charities and the University of California Board of Regents (my current employer).[ii] Even if many Californians today would be horrified at the idea of eugenics, there are some Americans and Californians now who are attracted to the idea of making better babies through genetics, and some of them are transhumanists. They embrace the idea of "positive" eugenics that focuses on making people "better." Where did this idea of "better babies" originate in America?

"Better babies" through eugenics

In the early 1900s, American schoolteacher Mary deGormo believed that she had to set high standards for both her students and later for human beings more generally. In order to promote these eugenic ideals, deGormo developed the novel idea of a "better baby" contest to be held at American state fairs.

This notion became a reality at the Louisiana State Fair in Shreveport in 1908 in the form of the "Scientific Baby Contest" held there [17]. Babies were "scientifically" evaluated at the contest by deGormo with the assistance of pediatrician Jacob Bodenheimer. The babies were graded based on physical measurements as well as on assessments of intelligence and other factors. At the next stall down from the babies, vegetables and livestock were probably evaluated as well for contests of their own. An odd juxtaposition.

Throughout the US, eugenicists started promoting such "better baby" contests. According to Martin Pernick in his article "Taking Better Baby Contests Seriously," eugenicists were pursuing certain ideals as a means to

---

[ii] http://historynewsnetwork.org/article/1796

promote "better humans" through controlled breeding that might be more reminiscent of agricultural practices [4, 5]:

> "Scientific agriculture was itself a new, contested development in early-20th-century America. University extension services, 4-H competitions, and the other agricultural models emulated by the better baby contests…"

In the US, 4-H is an American agricultural, youth development, and mentoring organization with the four H's standing for the following areas: head, heart, hands, and health.

Better baby contests went on for decades.

At the 1931 Indiana State Fair, a Better Baby Contest was held that was captured on film (Figure 7.2).[iii] By that time, eugenics had already been a popular notion in the state for many years. Indiana had instituted the first law for sterilization of "inferior" people in the world in 1907. Astonishingly this state law and then similar laws (the original was revoked, but a new law was passed later) stayed on the books in that state until 1974.[iv] This led to approximately 2,500 governmentally forced sterilizations. The poor, uneducated, people of color, Native Americans, and people with disabilities were disproportionately targeted.

The Better Baby Contest in 1931 in Indiana reportedly favored "educated white families and their ideally robust children" and such "ideas of biological betterment found uniquely fertile soil in the Hoosier Heartland." Residents of Indiana are sometimes called "Hoosiers."

According to a historical article from Indiana University,[v] eugenicists in Indiana advocated for forced sterilization and the Indiana "Committee on Mental Defectives" surveyed the state for candidates for "intervention" amongst the poor:

> "In the late 1910s and early 1920s, the Indiana Committee on Mental Defectives — an organization based on eugenic theory — sent women field workers across the state examining the living conditions and behaviors of

---

iii http://newsinfo.iu.edu/news/page/normal/15631.html
iv http://www.uvm.edu/~lkaelber/eugenics/IN/IN.html
v http://newsinfo.iu.edu/news-archive/15631.html

Figure 7.2. Part of a photograph from the 1931 Indiana State Fair Better Baby Contest. Image credit: Indiana State Archives. Reproduced with permission.

Indiana's poorest citizens…moving state policy away from mere condemnation of the poor to an approach which favored sympathetic intervention."

When I think of competitions at state fairs, such as here in California today, I imagine contests for the biggest fruits or vegetables. Students belonging to 4-H programs might participate. Perhaps also the best or biggest farm animals such as pigs are on display with blue ribbons.

In Indiana, more than 80 years ago, the same idea was applied to human babies. At the 1931 Indiana State Fair, scorecards were handed out to the doctors and others examining babies. The judges entered scores based on listed criteria (Figure 7.3). The best score for a "better baby" was 1000 points.

On this particular scorecard that I have included here from the Indiana Historical Society, someone has written "Defects or diseases not

Better Babies Standard Score Card

Entry No._____ Division _____
Score _____ Age in Months _____
Rural _____ City _____
Male _____ Female _____
Name _____
City and State _____
Street and Number _____
Weight at birth _____ lbs.
Strong or weak at birth _____
1st, 2d, 3d, 4th, 5th, 6th, 7th, 8th, 9th, 10th child _____
Breast-fed _____ No. months _____
{ Mixed-fed (bottle and breast) _____
{ No. months _____
{ Bottle-fed _____ How many months _____
{ What foods _____
Amount of milk in each feeding _____
Number of feedings now in 24 hours _____
Kind of food at present _____
Sleeps alone _____
If not, with whom _____
Sleeps in open air _____
Windows open _____ How many _____
Father's name _____
Age _____ Nationality _____
Occupation _____
Mother's maiden name _____
Age _____ Nationality _____
Occupation _____
Has birth been registered _____
Where _____
Contest held at _____
By _____
Date _____

ISSUED BY THE
BETTER BABIES BUREAU
WOMAN'S HOME COMPANION
381 Fourth Avenue, New York
Copyright, 1913, 1914, by The Crowell Publishing Company

Test IV—Oral and Dental Examination

If possible this examination should be made by a dentist.
See leaflet "Suggestions to Physicians."

	Maximum Score	No. Points Deducted for Defects
MOUTH. Habitually held open (10)___gums unhealthy, or abnormality (5)___tongue coated (5)___protruding (5)___hard palate (high arch or other deformity) (10)___uncleanliness (5)___deformity caused by thumb-sucking (5)___	45	
Lips: poor color (5)___fissures (5)___	10	
TEETH. Number of teeth___delayed teething (5)___	5	
___Prolonged retention of deciduous teeth (10)___supernumerary teeth (5)___	15	
Discolorations (5)___decayed teeth (5) (10)___irregular teeth (10)___notches or ridges (5)___	30	
Malocclusion (10)___	10	
Maximum total score___	115	

Actual total score in Test IV _____

Examiner _____

Dentition Table

Lower central incisors	at 7th mo.–Interval 2 to 3 mo.
Upper central and lateral incisors	at 9th mo.–Interval 2 mo.
Lower lateral incisors	at 12th mo.–Interval 2 mo.
First molars	at 14th mo.–Interval 2 to 5 mo.
Cuspids	at 18th mo.–Interval 3 to 5 mo.
Second molars	at 26th mo.–Interval 3 to 5 mo.

Test V—Eye, Ear, Nose, and Throat

If possible this examination should be made by a specialist.

	Maximum Score	No. Points Deducted for Defects
EYES. Abnormalities of size (5)___position (5)___condition of conjunctiva (5)___discharge (5)___cross-eyes (5) (10)___lids (5)___	35	
EARS. Abnormalities of size (5)___shape (5)___	10	
NOSE. Obstruction in breathing with mouth closed (10)___discharge (5)___offensive breath (5)___	20	
THROAT. Enlarged or diseased tonsils (10)___adenoids (10)___	20	
Maximum total score___	85	

Actual total score in Test V _____

Examiner _____

Maximum Total Score in all Five Tests...1000
Total Number Points Actually Scored in all Five Tests..........................

To ascertain actual score add together number of points scored in all five tests. To reduce score to percentage place decimal point before last figure.

Figure 7.3. Better baby score card. Image credit: Indiana State Archives. Reproduced with permission.

listed" so that apparently the doctors or others scoring the babies could add other factors into the scoring. Also included elsewhere on the scorecard were suggestions for those interested in organizing Better Baby Contests of their own.

Reportedly, leading eugenicist Charles Davenport became aware of the better baby contests and suggested to some of the organizers that

Figure 7.4. Fitter Family contest medal awarded the American Eugenics Society, 1920s. Source: Center for Genetics and Society used with their permission.

heredity must be included as well [6, 7]. These better baby contests eventually also spawned a second, related entity: fitter family contests.[vi] The fitter family contests also took parental qualities into consideration. Davenport recommended that 50% of the score should be for heredity, saying to Iowa better baby organizers about their imperfect contests that, "a prize winner at two may be an epileptic at ten." Proponents of fitter family contests included Dr. Florence Sherbon and Mary T. Watts, who held such competitions in Iowa after first holding better baby contests as early as 1911. Winners would receive prizes including medals (Figure 7.4). There is a sense that as the better baby contests evolved into fitter family competitions, eugenic principles more strongly took hold.

These contests resonate today as we think about possible efforts at human betterment through genetic enhancement. Could new genetic technology push us in the direction of a world more like that with "better baby" and "fitter family" contests? Some favor the idea of striving towards making "better babies" today and in the future.

[vi] http://bit.ly/18olVJd

Oxford bioethicist Julian Savulescu debated Robert Sparrow on the contended pros and cons of "making better babies" in 2012 based on new technologies not available during the early days of American eugenics [8]. I discuss this debate further at the end of this chapter.

Duke Professor of Law and member of the US Presidential Bioethics Commission Nita Farahany also has favored allowing human enhancement. Over the years, she has advocated in particular for allowing three-person IVF and possibly other genetic modifications of humans in specific debates. For instance, her words at a debate[vii] over whether to prohibit production of designer babies (she was against the prohibition) strongly resonate:

"…the enhancement versus therapy distinction is really just a red herring. So do parents want to enhance their children? Yes. Do they already do it? Yes. Should genetic engineering that enhances their health be permitted to go forward? Yes."

You can watch video of another debate where Professor Farahany expressed similar views.[viii] I also talked with Farahany in an interview that I posted on my blog,[ix] where she sounded more cautious. She was supportive of three-person IVF but opposed human embryo editing; "I'm in favor of mitochondrial transfer, but not nuclear gene editing at this time. We haven't reached a point in the technology where nuclear gene editing could be done with an expectation of safety and efficacy." However, she left the door open to the possibility in the future of her supporting designer babies with germline edits if safety and efficacy could be established.

By nature, most of us parents want our children to be healthy and happy. One could view basic parenting efforts as a form of "enhancement" over the grim alternative of putting your child at risk of malnutrition and such. However, common sense dictates that doing things such as feeding our child a healthy diet and taking care of one's own health as a

---

[vii] http://bit.ly/1GuP3t8
[viii] http://bit.ly/1frZ3wN
[ix] http://www.ipscell.com/2015/07/farahany/

mother during pregnancy are entirely different than genetically enhancing your child by heritably altering her or his DNA in every cell of their body.

Well beyond California and Indiana, other US states including Oregon experienced what can accurately be called eugenics-based disasters involving forced sterilization. Overall in the US eugenics laws, again some continuing as late as into the 1970s, created an enormous tragedy. Greater than 60,000 people were forcibly, but legally sterilized by governmental agencies in the US. While some states such as Virginia and North Carolina have acknowledged this wrong and sometimes even made monetary reparations, overall there is a troubling historical amnesia about American eugenics.

It may be hard to imagine today, but it is possible in the future that a similar, but inverse type of tragedy could occur based upon forcible genetic modification. This disturbing possibility is discussed further later in this chapter. For more on eugenics in America I recommend the fantastic book, *Eugenic Nation: Faults and Frontiers of Better Breeding in Modern America* [9] by Alexandra Minna Stern.

IVF and eugenics

Eugenics was popular during the 20th century in Britain as well. For example, fertility innovator Robert Edwards was also a eugenicist. Transhumanism, the movement based on the belief that we should push to transform humanity into something fundamentally better, was also emerging.

Edwards was asked about eugenics during a hearing conducted by the UK House of Commons' Science and Technology Committee. He made this statement in response to a question from a member of the UK parliament, Dr. Brian Iddon:[x]

"Again, it depends what we mean by eugenics...the word became degraded and it is a word that you have to be very careful about using

---

[x] http://bit.ly/1DZxnTH

today ... I think we can define what we want to do without using that term and I think we can make it clear to people what we want to do without using that term. I would love to show you the minutes of our meeting last week when these things were discussed in great detail when you would have seen that the problem is complicated but could be solved."

Eugenicists of the type of Edwards, and some (but not all) of today's transhumanists who have eugenics-like views, adopt more of the "positive" eugenics perspective. This means that while they do not advocate forced sterilization or controlled breeding, they do support human genetic enhancements. It is more about making people "better" than restricting people who might be perceived as inferior.

Edwards was not just enamored with eugenics in a passing phase. He reportedly became a long-time proponent of the eugenics movement in the UK. He was a leading member of the Eugenics Society and served for many years on its Council.[xi] The Eugenics Society still exists to this day, but has renamed itself The Galton Institute (apparently no longer a eugenics-promoting institute today) after the founder of the eugenics movement, Sir Francis Galton,[xii] who coined the term "eugenics." This term literally means something akin to "well-born," "good race," or "good stock." You can read much more about the wild life of Galton as well as his powerful influence on society in Nathaniel Comfort's intriguing book *The Science of Human Perfection* [10]. Comfort points out that Galton and eugenics were so popular that Galton was able to establish an endowed professorship of eugenics at University College, London.

Edwards, from a eugenics perspective, perceived links between human "quality" and genetics, and was not afraid to tie the two together:[xiii]

"Soon it will be a sin of parents to have a child that carries the heavy burden of genetic disease. We are entering a world where we have to consider the quality of our children."

[xi] http://bit.ly/1F6sKN9
[xii] http://www.galtoninstitute.org.uk/about.htm
[xiii] http://bit.ly/1zP58Mt

It is not difficult then to imagine a future in which gene-editing technologies empower a new eugenics. A more powerful form of eugenics boosted by cutting edge technology could pose major risks to society.

What is a "better" or "perfect" person?

If the various tools and resources come together to allow us to design our babies in the future with the goal of somehow improving or even perfecting them, a logical question would follow: what would constitute a "better" baby or a "perfect" person?

In Glenn McGee's book *The Perfect Baby* [11], he describes the "millennium-style, consumer-culture" conception of a perfect man:

> "6 feet tall, weighing in at 185 pounds, without hereditary disease. His brain is engineered to an IQ of 150, with special aptitudes in biomedical sciences. He has blond hair, blue eyes, archetypal beauty, and poise. Neurotic and addictive tendencies have been engineered out, as has any criminal urge, but in the male model, aggressiveness is retained as part of the 'athleticism' package: muscular and quick, he is competitive and can play professional-level basketball, football, and hockey. He has the 'sensitivity' package and enjoys poetry from several cultures and periods."

There are a number of difficulties with this idealized societal notion of "perfection," as I am sure McGee is keenly aware. It is, for instance, Aryan-centric in terms of physical attributes, something out of Nazi Germany. McGee does not include in this profile certain other traits in the "perfect man" that may prove desirable to some: a high sperm count, no male pattern baldness, hard teeth that do not get cavities, resilient skin of a certain pigmentation, eidetic memory, creativity, and wisdom.

Still, McGee's description is powerful and resonates intentionally, I imagine, with a commercial undertone. There is also the appealing, but unrealistic notion of separable "packages" of attributes that can be added in a modular manner. This does not reflect the reality of biological complexity. For example, aggressiveness associated with somewhat higher testosterone levels may predispose a male to certain violent and criminal tendencies.

Getting back to intelligence, why should perfection stop at an IQ of 150? Why not go even higher to the super-genius level? What about 160 or 170...200? Most likely the reason is that such an extraordinarily high IQ would often be accompanied by other traits that may be viewed as undesirable, such as depression, obsessiveness, or megalomania.

Other countries, cultures, and times in history or in the future might perceive "perfection" in quite different ways and characterize it by very distinct specific physical and psychological traits. Human perfection will also always be in the "eyes of the beholder."

Let us discuss a hypothetical GMO sapiens named "Pat Perfect" to illustrate the complexities of the quest for human perfection and the very different possible perspectives. Pat Perfect could be female or male, but in this example we will say Pat is female.

From an individual's perspective such as Pat Perfect, her own sense of perfection will center on what makes her feel good about herself and her life: self-confidence, sense of self-satisfaction, accomplishment, and longevity. The way Pat Perfect sees perfection in her own traits might clash with the views of others around her, and might even be terribly problematic for them.

For instance, Pat Perfect's children might not view her in this incarnation as so perfect. These kids could well view the very same traits that Pat Perfect sees as "perfect" in a completely opposite way as negatives. For example, if Pat Perfect is a mega-genius wrapped up in her own self-satisfying intellectual world, she might be a terrible parent even if that same genius intellect makes her feel terrific about her own self. Further, if Pat Perfect is so absorbed in herself she could become narcissistic at the expense of her children. Pat's children's perceptions of perfection for Pat could center on her being a great parent instead.

From strictly a parent's perspective, our GM "Pat Perfect" could end up being designed and produced very differently than if Pat Perfect herself or her own children were in the driver's seat of genetic design. A perfect child from the parental view such as Pat might be one that makes a parent feel good about themselves, perhaps even mirroring certain aspects of the parent even if those parental traits are questionable. There is a certain level of vanity or narcissism in all of us. Such a parent's "perfect child" might be relatively quiet so as not to bother the parent. The parent might want Pat Perfect to be a high performer in school, sports, and at the workplace to boost parental self-worth. Pat's mom and dad might want her to possess high fertility to provide grandchildren, and good parenting skills so as to keep the grandchildren doing well.

From a societal perspective, human "perfection" is likely to be very different to that of the parental, children's, or individual ideals of perfection. Mental and physical health could be preeminent for Pat Perfect based on societal priorities. A predilection to not question authority might be prized. Very high or low intelligence would be less than ideal. People who are "too" smart could pose risks to society, while people who are exceptionally lacking in intelligence could cause trouble or drain resources. Longevity might not be a priority and might be "discouraged" genetically because of health care costs. On the other hand, as long as it was both a long and healthy life, from society's view Pat Perfect could aid humanity by living a long time and passing along wisdom and knowledge to those around her.

Other conflicts may arise as well over different perceptions of perfection and priorities. For example, a certain couple might consider their new child higher on the scale of perfection if it was male, but from a societal perspective it could be preferable if the child were female based on present population dynamics (e.g. in China and India today there are literally too many males relative to females due to sex selection). Or vice versa. Sex selection via PGD is already growing in use and could become even more common.

What might result from such tensions between distinct views of "perfection" if we start genetically tinkering with our children? There could well be powerful consequences, such as conflict, social dysfunction, and even violence.

Should we try to make "better babies" via genetic modification?

As briefly stated earlier, in 2012 professors Julian Savulescu and Robert Sparrow held a spirited debate on the pros and cons of using genetic technology to attempt to make "Better babies." Savulescu was on the pro side and Sparrow argued that such efforts to make better babies would be misguided. I describe both perspectives from that debate below, starting first with the pro side articulated by Savulescu.

The "pro" arguments. Savulescu started off by arguing that the selection of human reproductive cells (also called gametes) with the intention of making better babies is already widespread and pointed to sperm banks and oocyte donation as examples. Those now seeking to purchase gametes can and do make selections based on any number of donor attributes: intelligence, education, height, pigmentation, and more. He is right about that.

At the moment, when parents-to-be select donated eggs or sperm on the basis of specific traits, they do so in the hope that their resulting offspring will inherit such "better" traits and physical features similar to them as the parents-to-be. This is arguably a simplified form of the 23andMe "build-a-baby" tool, which I discuss in Chapter 5, and of the GenePeeks tool.

Sperm and egg selection is very real and could well be termed as an effort to create "better babies." Savulescu wants to establish that today's reality already allows for this certain level of human design, or as he puts it, "Despite the rhetoric, in practice we love designer babies." My sense is that "designer babies" should be a term that is restricted to referring to babies that specifically have specific laboratory-introduced genetic modifications.

With this foundation — that pursuit of "better babies" is already occurring in our world today — Savulescu goes on in the debate to pitch the notion that designing better babies is not just happening and is okay, but also is outright prudent and should be encouraged.

He argues for a form of "genetic determinism." The same way that some have said, "anatomy is destiny," others like Savulescu might argue that "genetics is destiny." He also notes that designing organisms with genetics can be successful and uses dogs as an example:

> "The three hundred different breeds of dogs that are around today are all the result of genetic selection over ten thousand years. Some are smart, some are stupid, some are vicious, some are placid, some are hardworking, some are lazy, that's all genetic. No matter how you treat or train a Chihuahua, how many vitamin supplements you give it, it will never beat a Doberman in a fight."

In principle, Savulescu says that what works for dog breeding should work for humans as well, except hugely accelerated by genetic technology. He follows up by saying, "What took us ten thousand years in the case of dogs could take us a single generation through genetic selection of embryos." Make no mistake, he means human embryos.

I suppose one might follow up by asking parents-to-be whether they would prefer that their children be Chihuahua or Doberman type of people? If those in your peer group and our social competitors are choosing Doberman-style human children, are you really going to choose to have a Chihuahua child?

In his book *Who's Afraid of Human Cloning* (Page 168) Gregory Pence, Professor of Philosophy at University of Alabama at Birmingham, also brings in the disturbing dog-human trait modification analogy [12]:

> "Many people love their retrievers and their sunny dispositions around children and adults. Could people be chosen in the same way? Would it be so terrible to allow parents to at least aim for a certain type, in the same way that great breeders…try to match a breed of dog to the needs of a family?"

I find the idea of creating "sunny" children via genetics to be creepy.

Building on the notions that we are already making designer babies (seeking to normalize this practice in his argument) and that genetics works powerfully to influence animal traits, Savulescu then asserts that human traits are equally important as human health. If we can use genetic technology to improve human health, then why not use it to "improve" human traits? He indicates that some human traits are definitely superior too and should be valued.

He circles back to the notion of genetic determinism again and falls into this idea, propagated by some transhumanists as well, that single genes might be targeted to confer certain "better" traits. It is an attractive idea to some, but lessons from genetics research suggest it is likely to be in most cases an overly simplistic notion.

He points to several gene variants that are associated with conceivably desirable behavior: a variant of the *COMT* gene that is associated with altruism; an *MAOA* gene variant associated with non-violence; and a mutation in the *AVPR1A* gene linked to fidelity. He seems to suggest that we should focus on these "better" genes to yield "better" babies. Going further, he argues that we have a moral obligation to do this. However, genetics does not work in this straightforward kind of manner in reality.

Again, in an attempt to normalize his rather extreme notions about genetic and reproductive practices (which sure sounds like eugenics), he compares such efforts at genetic trait improvement to other ways that parents actively work and have an obligation to make things better for their children: a good diet, excellent education, and avoiding poverty.

Savulescu is advocating a new eugenics. He emphasizes that in this technologically driven "liberal" eugenics, things would be different than in the past disastrous efforts at eugenics:

> What differentiates the so called new liberal eugenics from the old Nazi eugenics are two things. Firstly, people are free to make their own choices and decisions — and even to refuse to have genetic testing. Secondly, it is not based on race or on the Social Darwinist values that the Nazi program was.

If gene editing-based human genetic modification becomes widely available, it is reasonable to think that many parents-to-be (especially if they are economically well off) will use this more powerful technology, not only to influence the health of their future babies, but also their traits as well. Savulescu argues this would be a good thing.

The "con" arguments: Professor Sparrow has a very different view of both PGD and efforts to make "better babies." Even so, he agrees with Savulescu that these technologies might in some cases be ethical to use to prevent defined health problems. Where they diverge most strongly is over the use of these technologies to generate enhanced babies with "better traits." Sparrow is intensely opposed to this.

He asserts that such an effort could become a never-ending quest for some failed notion of perfection or going for the "best," not just better. If you generate a longer-lived human via the PGD-based choice of an embryo with gene variants that associate with longevity, what is to say you did not fail to select an even longer-lived one? At what point would the baby or human produced satisfy this desire for the "best" in terms of longevity and fighting aging? For instance, why stop at an average age of 100 years for your GMO sapiens if you could shoot for 120 years? Then 140 years and so on. It is a never-ending quest, Sparrow says.

For traits more generally, a new GM or PGD-selected baby could always be generated who grows up to be longer-lived, more intelligent, more attractive and so forth than the choice that at some point you "settled for" as a parent. There is also a sense amongst proponents of designer babies that today's current human traits are not good enough.

If we really value producing the best possible children, Sparrow argues, then perhaps we should not even be having our own genetically-related babies, because surely some people out there have better genes and hence better traits than us. It is in a sense an extreme notion, but Sparrow is rightly making the point that the pursuit of "betterness" through genetics can logically be extended to dangerous extremes. Sparrow also is concerned about the social outcomes that could be catalyzed by efforts to produce "better" children:

> "As Julian has observed, I do worry that in sexist or racist societies the best child — the child with the highest expected welfare — will be a blonde-haired, blue-eyed, male child. The impact of racism and sexism on the welfare of children is very real. It is as real as many other things that we could influence genetically."

I share these concerns. I also worry that genetic-assisted campaigns to generate "better" children would lead to decreased diversity in our species and to more discrimination against certain classes of people. If we have a duty as parents to teach our children to value diversity and to fight racism and sexism in their daily lives, do we not have an obligation as parents to avoid letting racism, sexism, and other forms of discrimination influence our reproductive choices?

Sparrow finally argues against genetic efforts at designer babies by invoking a potential dystopian new reality in the future caused by such efforts. In this reality, parents may be forced to "choose" designer babies based on social and governmental pressures.

In the end, both Savulescu and Sparrow raised intriguing points regarding eugenics and designer baby technology. One of the most powerful concepts to emerge from their debate is that "better baby" technology could well prove to be unpredictable and uncontrollable, shifting at the whims of societal preferences and trends.

Transhumanism: Getting from point ACGT to H+

Is humanity already getting better over the years, centuries, or millennia? Is each of us as an individual better than human beings of the past and will humans in the future be better than us?

Could religion be the key to human improvement? Might better education pave the way? Or is science the real ticket to human betterment?

Believers in the transhumanism movement advocate for human transcendence to become a new and better species: *Homo evolutis*. Transhumanism today has embraced new technologies, including genetics and genomics, as a means for this transcendence. They also find value in other technological efforts, such as establishing human colonies off of Earth, moving human consciousness into machine form, and more.

For over 50 years, transhumanism has advocated the notion that human improvement is desirable and that such improvement should be of a grand scale. They want really big changes. Ideally, we humans should become so much better that we become transhuman, meaning a new species. From my view, transhumanists are some of the most intelligent, creative, and accomplished people in the world. I do worry about how some of them appear to advocate for human genetic enhancement though and also how their notion of "betterness" could lead to societal conflict.

Even though the key to transhumanism is the role of science in human improvement rather than that of God or religion, transhumanism has a distinctly religious feel to it. For example, when we become something better than human, there is a sense of glory to that. Because of the overriding emphasis placed on betterment, the movement refers to itself by the moniker H+ or h+ (Figure 7.5). The "+" along with the letter h are meant to symbolize going beyond human.

Figure 7.5. The h+ symbol of transhumanism. Image source: Wikimedia.

Genomics and gene-editing technologies are a means for, a transhumanist might say, getting from point A (or ACGT to put a DNA spin on it) of today's undesirable human reality to point h+ where we are far better. Could genetics get us to that point of a new, better h+ being? Some advocate this path.

Early 20th century philosopher Julian Huxley, often touted as the founder of transhumanism, coined the term "transhumanism" in an essay in 1957 going by the same name.[xiv]

Huxley wrote that humans are in the driver's seat when it comes to evolution:

> "It is as if man had been suddenly appointed managing director of the biggest business of all, the business of evolution — appointed without being asked if he wanted it, and without proper warning and preparation. What is more, he can't refuse the job. Whether he wants to or not, whether he is conscious of what he is doing or not, he is in point of fact determining the future direction of evolution on this earth. That is his inescapable destiny…"

What more effective way to take control of human evolution than through human germline genetic modification? Is the making of GMO

[xiv] "New Bottles for New Wine" by Julian Huxley, 1957.

sapiens the only way to get to the transhumanist *Homo evolutis*? To have a lasting impact, human "betterment" would need to be hardwired into us and persist despite changes in our world.

In theory, if you buy the argument that humanity and humans have gotten better over the last thousand years or even only a hundred years, consider how much of that change for the better could be erased rather quickly. With world war, famine, plague, or some other catastrophe that wiped out our technologies and our libraries of information of one kind or another (books, computer databases, etc.), would the remaining human survivors really be better than those from a few thousand years ago? Perhaps not. For real and lasting transhumanist change humans need to be coded differently.

Today's ongoing gene therapy trials can make individuals "better" in the sense of resisting or overcoming a disease, but once that individual dies, the change is gone. In contrast, genetic modifications made to the reproductive cells of humans could make persistent changes in human beings that are inherited from one generation to next. These changes would continue to be inherited even if our knowledge base was lost. Sound like sci-fi? The technologies that are necessary to make h+ humans with heritable genetic changes are upon us now. While to my knowledge no one has done this yet, it could be coming soon.

What does "better than human" or h+ really mean?

More disease resistant? Smarter? Better parent? Kinder? Wiser? Can we genetically modify humans to attain such goals? And can those new traits conveying the new, better phenotypes be inherited?

Some might say that transhumanism fueled by genetics technology poses transcendent risks as well. What if your smarter h+ human is inclined to mental illness? There is some evidence that intelligence correlates with mental illness. What if your h+ human that can never get Alzheimer's or autism is much less kind than the average person today? Perhaps even inclined to cruelty? These are just two examples.

One problem with the idea of transhumanism where it links up specifically with genetic modification is again the reality that complex traits are not generally controlled by one gene (what scientists call, "monogenic.") Similarly, it is arguably a gross oversimplification to think that by introducing a gene edit into the human germ line that there will only be one consequence and that one being just the desired change for the better.

Transhumanism is alive and well today. There is even a society for transhumanism called Humanity+ and a magazine for transhumanists called "h+ magazine."[xv] In the magazine, there have been a number of articles on CRISPR-Cas9 technology, including one suggesting it might be used effectively to modulate humans to combat aging.[xvi]

Another article in this magazine, titled "Fix diseases and improve yourself through gene therapy," not only cites CRISPR-Cas9 as a possible tool for use in humans, but also suggests a genetically enhanced society would be a good thing.[xvii] This article even goes so far as to promote genetics-based cosmetic enhancement of people's appearance:

> "Since the concept of cheap and easy gene therapy to enhance one's looks or capabilities is so novel, most people have never properly thought about the issue before…There are many arguments in favour of gene therapy for the purpose of improving one's physical appearance. For example: It is safer and far more effective than plastic surgery, which a lot of people are already doing anyway…A society made of beautiful people will be more cheerful and pleasant for everyone, and people will not be judged solely on looks anymore, but more for their personality, interests, good actions, etc…One common objection to using genetic engineering to increase one's looks and intelligence is that people will end all looking the same and thinking alike. Nothing could be further from the truth."

Therefore the argument in this h+ article is that the genetic enhancement of looks would not make everyone look too similar, but where is the evidence to support such a prediction? We could not be sure one way or another what might happen if we go down this extreme path. I am not particularly optimistic about how it would turn out. Look at the lengths to which some people will go to look like their idols.

[xv] http://hplusmagazine.com/
[xvi] http://hplusmagazine.com/2014/09/17/crispr-future/
[xvii] http://bit.ly/1KUurOl

Personal enhancement via gene therapy or through parental enhancement of future offspring via genetic modification could well be subjected to the same kinds of broad trendy notions of what is "hot" and what is not. In an era of designer babies, those parents who choose that route could end up independently creating many children with the same or similar traits.

George Church, genetics pioneer and transhumanist

Do you still believe in Santa Claus or other kindly gift-giving magical beings? When it comes to genetics, I confess that I might. Harvard Professor George Church seems to me to be akin to something like a real Santa Claus of genetics. He even looks a bit like the part, with a bushy white beard, glasses, and a kindly face. In Figure 7.6, he is shown holding a molecular model that almost looks like a wrapped present to give to a child on a holiday. What is inside? The key to making us better?

Church also has an unquenchable enthusiasm for pushing genetic technologies to their very limits and beyond. In this way, he comes up with "presents under the tree" of innovations and ideas. I admire Church at least in part because he has a way of weaving what feel like out-of-this-world genetic innovations into fact-based realities. Still, we do not see eye-to-eye on heritable human modification. I am pessimistic about its possible consequences. On my blog, I often reach out for interviews with scientific innovators and in particular to those with whom I disagree on key issues.

In early 2015, I interviewed Church, perhaps the most famous of transhumanist scientists, for my blog (more on this interview in Chapter 9).[xviii] I have captured some of our interaction below, which resonated most strongly on the issue of transhumanism and genetic modification.

Knoepfler: Also they said this in the piece:[xix] "At meetings of groups of people known as 'transhumanists,' who are interested in next steps for human evolution, Church likes to show a slide on which he lists naturally occurring variants of around 10 genes that, when people are born with

---

[xviii] http://www.ipscell.com/2015/03/georgechurchinterview/
[xix] http://bit.ly/1aQ7QpI

Figure 7.6.   Professor George Church at TED 2010. Image credit: Steve Jurvetson.

them, give them extraordinary qualities or resistance to disease." Can you comment on that?

Church: Note that throwing in the word "transhumanists" is unnecessarily confusing. I use that slide at all sorts of meetings (none of them on "transhumanism"). There may be transhumanists in the audience, but that is not the point. The point is that: in addition to common variants of small impact and rare deleterious variants, there are rare protective gene variants of large impact.

Like Savulescu, Church points (Table 7.1) to specific human gene alleles associated with "better" traits in human beings. Could gene editing make these a reality in new human offspring? Would such designer babies really be better off? What about society as a whole?

In 2015, researches led by Professor Xi-Jun Yin created super-muscled GM pigs via editing one of Church's ten example genes, the

Table 7.1. List of 10 genes that George Church has cited as being linked to specific desirable traits that could be a focus of human genetic interventions.

Gene name	Gene varian	Associated trait
LRP5	G171V/+	Extra-strong bones
MSTN	–/–	Big, lean muscles
SCN9A	–/–	Insensitivity to pain
ABCC11	–/–	Low odor production
CCR5, FUT	–/–	Virus resistance
PCSK9	–/–	Low coronary disease
APP	A673I/+	Low alzheimer's risk
GHR, GH	–/–	Low cancer risk
SLC30A8	–/+	Low Type 2 Diabetes risk
IFIH1	E627X/+	Low Type 1 Diabetes risk

MSTN gene.[xx] The genetic modification was completed using the TALEN approach, but could have been done with CRISPR as well. These pigs, which the team called "double-muscled," would make body builders jealous with their massive muscles. More importantly if ever approved for human consumption these Arnold Schwarzeneggers of the porcine world (Figure 7.7) could produce more and potentially healthier meat. Keep in mind that no GM animal has ever received approval to be eaten by people. This research also raises the question of making GM humans with *MSTN* mutations with the goal toward enhancement via supersized muscles. The super-muscled pigs should be viewed as a cautionary tale because although 32 were created, only very few lived and just one of those GM pigs of this kind is entirely healthy. In addition, birthing problems were reported due to the large size of the GM piglets.

Juan Enriquez and Steve Gullans, the authors of a book entitled *Evolving Ourselves* [13], echoed some of the same pro-enhancement kind

---

[xx] http://bit.ly/1Kq1Cwb

Figure 7.7. "Double-muscled" pigs that have huge muscles due to an introduced genetic modification in the MSTN gene. Upper right insight is a picture of a normal pig for comparison. Image credit from Dr. Xi-Jun Yin, used with permission.

of sentiments as Church. They also point to specific desirable gene variants that are "better" than perhaps those in the DNA of most of us:

> Every year the genetic casino gets tipped a little more toward eliminating more disease carriers and inserting positive traits; perhaps inserting the *CETP* gene, associated with a 69 percent reduction in Alzheimer's. Or the *DEC2* gene so you only need six hours of sleep each night. A rare *APOC3* gene mutation may become a popular addition, as it's been found to lower fat in blood by 65 percent in the test population of Old Order Amish and to greatly lower Alzheimer's risk in Ashkenazi Jews. Japanese Americans who carry *FOX03A* have significantly lower rates of cancer and heart disease than the average American. Each of these discoveries increases the potential menu of desirable alterations to one's own genome, or that of one's baby.

This sounds like a possible recipe for designer babies. They rightly indicate that big ethical issues come along with gene-editing technology that could make clinical human modification a reality, but at the same time, they also correctly emphasize how simple it could be for people around the world to give it a try. They write with an infectious enthusiasm for human modification, a sentiment that I do not share.

Church also wrote a very stimulating book in 2012 called *Regenesis* [14], in which he articulates his vision for a technologically driven future, one involving synthetic biology and genetics. Gene editing technology is advancing at "warp speed." To my knowledge the words "CRISPR" and "Cas9," so dominant today in the dialogue of human germline modification, are not even in the index to Church's book published just three years ago. Where will we be three years now?

The last chapter of *Regenesis* is focused in part on designer organisms and human babies. In this chapter, Church and his co-author Ed Regis make the case that human genetic modification is something to move toward. They also emphasize, in an attempt at normalizing human genetic modification, that new forms of DNA that have never existed before on Earth are made routinely during human and other kinds of reproduction that occur naturally on our planet. To my way of thinking, this misses the point because designer babies would have focused, very specific, and deliberately engineered genomic changes toward a specific goal of "betterment." As I have discussed earlier that is not at all the same as normal human sexual reproduction.

They also acknowledge that there would be risks:

"… if genetically modified microbes pose a threat, what about genetically enhanced human beings, co-called transhumans? It's a cliché that the human being is the most dangerous animal. Would a race of transhumans be even worse?"

In a general sense, Church and Regis go on to answer this last question in the quote with a "No." My view is that such efforts could easily end up creating people with a host of health problems. They might be a danger to themselves and others.

Church and Regis rightly point to challenges in making designer babies. They cite, for example, the potential link between genius and madness or in some cases more moderate forms of idiosyncratic mental oddities. They also ponder the potential creation of an evil super genius whose malicious intentions could lead to large-scale disaster:

> "…how much worse would be the potential for damage wrought by an actively malicious supergenius? After all, the power of individuals is growing. In ancient times one person might murder a couple of enemies with a rock. Later, he could destroy a village with fire. Today a small team might kill millions of people with nuclear, chemical, or biological warfare agents. In the future, could a single transitional human having a bad day…release a doomsday WMD?"

In the end, Church appears more inclined to favor taking the risk for the potential benefits of creation of transhumans (what we might call GMO sapiens here in this book) via gene editing and designer babies, but at least with his eyes wide open and having given it deep thought. As humanity contemplates now possessing the powerful technology to change itself and its evolution through genetic modification, keeping our minds and eyes open will be key to doing what we can to avoid a resurgence of eugenics or other negative outcomes. I cannot wait to see what George Church does next even if I am also a bit unsettled by what could be coming.

Forced genetic change: gene drive, and weapons

Another concern with technology for making GMO sapiens is the notion of forced genetic change. We see this in the dystopian novel *Brave New World* and in other literature as well (discussed more in the next chapter). Eugenicists of the past went the other way — that of forced sterilization — to prevent reproduction and the passing on of "negative" genes, but could individuals or governments in the future force genetic modification upon us as well?

This is not something I have seen advocated by transhumanists, who go the other way to rightly value individual choice. Still, forced genetic change is possible. Some might see, for example, parents who choose to have a GMO sapiens child (particularly for a non-medical reason) as having forced that decision on the future child without their consent.

Beyond gene editing an individual person or other organism, another more powerful idea (both potentially for good and bad) is to catalyze specific genetic changes in an entire population or environmental system via GMO technology, including via gene-editing tools, such as CRISPR-Cas9. This hypothetical large-scale genetic process is referred to in a more general sense as "gene drive." Gene drive is defined as, "stimulating biased inheritance of particular genes to alter entire populations."[xxi] In other words, scientists could force genetic modification upon nearly an entire population of organisms via gene drive. In principle, these organisms could even be humans.

DNA cutting (aka "endonuclease")-based gene drive, which was proposed by evolutionary geneticist Austin Burt of Imperial College London, relies on the introduction of "selfish genes" that could be used as tools [15]. See my model of this technology in Figure 7.6. An endonuclease is an enzyme that cuts DNA or RNA; Cas9 is one such example. Selfish genes are genes that function in part to drive their own existence, potentially at the expense of other genes.

There is generally a 50% chance of any gene being inherited, but selfish genes break that rule and are inherited more often than that. An engineered selfish gene such as the one labeled "gd" in Figure 7.8, intended to promote gene drive could function by containing machinery such as CRISPR-Cas9 to cut the normal wild-type (wt) sister copy of the same gene. During DNA repair to fix the cut, many cells would use the mutant GM selfish gene as the template for repair. Two selfish genes are often the end result.

The goal of introducing gene editing-based gene drive, as articulated by George Church's group in a recent article [16], is to solve big challenges via technological innovations that involve forcibly introducing genetic modifications into organisms in the wild. The gene drive and its genetic elements would essentially take over the genetic state of organisms during

---

[xxi] http://bit.ly/1p0GFL9

the course of a few generations with the aim of achieving beneficial outcomes:

> "…gene drives could potentially prevent the spread of disease, support agriculture by reversing pesticide and herbicide resistance in insects and weeds, and control damaging invasive species."

The most commonly cited example of gene drive is to reduce malaria transmission by mosquitoes. What a great idea. Malaria kills millions of people. But you know there is a catch, right?

As attractive as GMO-based gene drive sounds, and I personally find it very intriguing, it is at the same time an extremely dangerous idea. The fundamental risk with GMO gene drive is that it may be difficult to control.

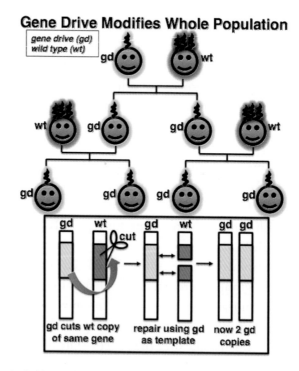

Figure 7.8. Author's model family tree of how gene drive can genetically modify and "take over" an entire population. The abbreviation "gd" stands for gene drive and "wt" stands for wild type.

Combine that with the real possibility that a GMO species could have unintended and very negative consequences in an environment once produced, and you can see how gene drive could make some very bad things happen.

Once released, the GM organism would not be controllable by us and hence if a negative outcome became apparent, there would probably be little if anything that we could do to stop it. A gene drive chain reaction could spiral out of control, permanently changing the world in significant and potentially negative ways.

In the GM mosquito case, a gene drive chain reaction could lead to ecological harm in the form of disruption of normal pollination patterns or changes in mosquito-prey or predator population dynamics. Because mosquitos also bite people and cause disease, it is also formally possible that GM mosquitos gone wrong could potentially cause new human diseases. It is a very unlikely possibility, but still a scary thought.

An actual GM mosquito experiment was just conducted in the wild in Panama in 2015. It was intended to reduce transmission of another mosquito-borne disease, dengue fever. This is a hugely important goal. The experiment involved releasing 4.2 million GM male mosquitoes that could breed with female disease-causing mosquitos in the wild, but would always lead to larval death due to an introduced genetic modification.

Although not gene drive, this was still an experiment with GM creatures conducted out of the laboratory in the "real world" intended to affect a wild population. Preliminary indications in the media (the work is not yet published so that is an important caveat)[xxii] are that the experiment reduced disease-carrying mosquito populations and no obvious negative outcomes were reported, although it might be too soon to know, and again there is no independent validation of the outcomes.

Gene drive and chain reactions based on it are also possible in the context of human beings, and there the situation could become even more concerning. GMO sapiens bearing gene drives could spread such genes through human populations over generations. The slow reproductive

---

[xxii] http://bit.ly/1clleCn

nature of humans would mean an equally slow spread amongst people, so hopefully we would see any problems unfolding before they get out of control, but maybe not. Even modified genes that have unpredicted negative outcomes, such as mental illness, could also provide selective reproductive advantages that facilitate their spread in humans, especially in the context of a gene drive gone wrong.

Innovative next-generation forms of CRISPR-Cas9 technology have emerged that could be particularly efficient at catalyzing gene drive. For example, in one new form of CRISPR-Cas9 technology, genes are altered in such a way that (in addition to the desired gene edit) the actual CRISPR-Cas9 machinery itself is also inserted into the organism's genome [18].

So-called "accidental release" is another potential problem with gene drive experiments, where GMOs intended to stimulate gene drive only in a laboratory setting are inadvertently released into the outside environment. Professor Burt discusses the risk, in a non-human context, at the end of his article:

> "In carrying out this work, it should be noted that the ease and rapidity with which these selfish genes can invade a population applies not just to planned releases, but also to unintentional releases of laboratory escapees. Proper attention to containment issues is needed to prevent natural populations of D. melanogaster, to take an obvious example, from being infected (and possibly endangered) by engineered selfish genes that were meant to stay in the laboratory. Finally, wide-ranging discussions are needed on the criteria for deciding whether to eradicate or genetically engineer an entire species."

GM humans could spread gene drive via their own reproduction, just as well as GM fruit flies made in a laboratory could do it if they escape the laboratory. GMO sapiens embryos created in a laboratory solely for experimental purposes could be at risk of intentional "release" if someone misappropriated them and implanted them into surrogate mothers. As pointed out by leading geneticist Harmit Malik of the Fred Hutchinson Cancer Research Center in a conversation I had with him on gene drive, it could spread in an insidious manner without us even being aware of it until too late:

Knoepfler: What about if we have researchers who conduct a real world gene drive experiment. What could go wrong?

Malik: Horizontal transfer is possible where the gene drive gets vectored into something else so in that case you end up with another organism having the gene drive with unknown consequences. It could be very insidious. Another issue that may arise relates to possible changes in the gene drive itself as it spreads through a population. Parts of the gene drive could mutate. In particular if the guide sequence mutates, then the gene drive could end up hitting some other gene and lead to unintended consequences.[xxiii]

In their article on gene drive, Church and colleagues briefly outline a possible contingency plan should something go wrong with a gene drive effort. One such approach is to take yet another stab at gene drive in an effort to reverse the first deleterious gene drive. It is unclear how well such a rescue effort might work and whether this sort of attempt at a "reverse gene drive" to correct the original gene drive gone wrong could cause even more problems in the real world. In our interview, Professor Malik was skeptical that reverse gene drive would work to effectively fix a gene drive problem.

Another idea to reduce risk from gene drives would be to include a second component in GMOs in the form of an inducible suicide gene or a unique fatal weakness to a synthetic chemical not present in the environment. For instance, with a GM mosquito or other insect that hypothetically causes unexpected environmental harm or moves to an unintended location, it might be possible (if a safety gene was included during their creation via gene editing) to kill off all the GM mosquitos with spray of a certain insecticide to which they were engineered to be susceptible. Of course dousing an environment with such chemicals

---

[xxiii] http://bit.ly/1QgHRXB

could have its own negative consequences beyond the target GMOs. Researchers at MIT have developed a so-called "kill switch" for GMOs to limit damage they might cause in the wild.[xxiv]

More broadly, should negative outcomes related to gene drive occur from human germline gene editing, the killing or sterilization of the GMO humans would, hopefully, be considered unacceptable. One of the most likely governmental reactions would still be quite disturbing though in the form of "reproductive quarantine." In this scenario, GMO sapiens who have traits in their germlines that are perceived as "negative" would be prevented by their government from reproducing. This would raise very serious ethical and societal issues, and pose a threat to democracy.

Even if specific human genetic modifications are widely seen as unwise in hindsight, other parents might still choose, nevertheless, to make similar GM choices in the future for their children despite the red flags already raised. In a worse case scenario, we might come to experience social pressure or even governmental mandates for us to genetically modify our children to, for example, reduce health care costs.

Another chilling possibility also comes to mind. Gene drive technology designed for the purpose of human genetic modification and combined with viral technology could be used as a weapon. Exposing human populations to a highly infectious agent that causes genetic modification and has a gene drive component would be a terrorist act that harms humanity. It could spread throughout the world like a plague because of the gene drive component and the transmission component of a highly infectious virus. Even the threat of the use of such a GMO sapiens-inducing weapon could have severe consequences. If people were intentionally or accidentally exposed to a genetic modification virus with a gene drive component, mandatory human reproductive quarantine may become a necessary evil.

Given the great power of genetic modification technology to do both seemingly very positive and terrible things in the future, it will need multiple levels of regulation and oversight. Ethicists are starting to weigh in more often on this issue as well. For example, bioethicist Jeantine

---

[xxiv] http://bit.ly/1GKyXiM (free registration required)

Lunshof in a May 2015 piece in *Nature* called "Regulate gene editing in wild animals"[xxv] outlined the risks of gene editing as well as gene drive in animals and explained an urgent need for regulation:

> "Once introduced, these genetic changes are self-propagating. If released beyond the laboratory, the effects would spread with every new generation and would quickly run out of control."

Keep in mind that her concern is focused on this happening in GM animals, but what about in humans? Surprisingly, she specifically expressed less concern about human genetic modification and even downplayed concerns over potential designer babies.

How we all make decisions about this area of science will have a profound impact on our species and our world. Because of the heritable nature of some forms of genetic modification, our decisions have importance beyond today and extend to the future.

Another serious concern about eugenics, transhumanism, and human modification more broadly is the possibility of reducing genetic diversity. One might imagine that if certain genetic modifications become trendy amongst parents-to-be that over time genetic diversity of certain human genes might dwindle. Genetic modification technology could empower parents to attempt to make their children have traits in common with celebrities. Recall California Cryobank's celebrity look-a-like sperm donor tool discussed in Chapter 6.

In a transhumanist H+ future, a new organization inspired by the American agricultural youth program 4-H might arise called 4-H+. If you recall, the areas of focus of the current 4-H organization are head, heart, hands, and health. In a transhumanist 4-H+ group the four key areas might be the following "better" elements: heads above, heartier, handsomer, and healthier.

In a transhumanist future, the reality is that only certain people would have the opportunity to become H+ and socioeconomic class would have a lot to do with that. Many of the rest of us would be left behind as simply an "H" rather than an H+, or a plain *Homo sapiens*

[xxv] http://bit.ly/1PFqNsr

instead of a wonderful *Homo evolutis* or *GMO sapiens.* Beyond transhumanists and eugenicists, how does the wider culture view human genetic modification? How is it portrayed in art and literature? Are there lessons there? I address these important questions in the next chapter.

References

1. Osborn, HF (1921) The Second International Congress of Eugenics Address of Welcome. *Science.* **54**:1397:311–3.
2. Farber, SA (2008) U.S. scientists' role in the eugenics movement (1907–1939): a contemporary biologist's perspective. *Zebrafish.* **5**:4:243–5.
3. Black, E (2003) *War Against the Weak: Eugenics and America's Campaign to Create a Master Race.* Four Walls Eight Windows.
4. Pernick, MS (2002) Taking better baby contests seriously. *Am J Public Health.* **92**:5:707–8.
5. Stern, AM (2002) Making better babies: public health and race betterment in Indiana, 1920–1935. *Am J Public Health.* **92**:5:742–52.
6. Lovett, LL (2007) Fitter families for future firesides: Florence Sherbon and popular eugenics. *Public Hist.* **29**:3:69–85.
7. Selden, S (2005) Transforming Better Babies into Fitter Families: archival resources and the history of American eugenics movement, 1908–1930. *Proc Am Philos Soc.* **149**:2:199–225.
8. Savulescu, JR Sparrow (2013) Making better babies: pro and con. *Monash Bioeth Rev.* **31**:1:36–59.
9. Stern, A (2005) *Eugenic Nation: Faults and Frontiers of Better Breeding in Modern America.* University of California Press.
10. Comfort, NC (2012) *The Science of Human Perfection: How Genes Became the Heart of American Medicine.* Yale University Press.
11. McGee, G (1997) *The Perfect Baby: A Pragmatic Approach to Genetics.* Rowman & Littlefield Publishers.
12. Pence, GE (1998) *Who's Afraid of Human Cloning?* Rowman & Littlefield.
13. Enriquez, JS, Gullans, S (2015) *Evolving Ourselves: How Unnatural Selection and Nonrandom Mutation are Changing Life on Earth.*
14. Church, GME Regis (2012) *Regenesis: How Synthetic Biology will Reinvent Nature and Ourselves.* Basic Books.
15. Burt, A (2003) Site-specific selfish genes as tools for the control and genetic engineering of natural populations. *Proc Biol Sci.* **270**:1518:921–8.

16. Esvelt, KM, *et al.* (2014) Concerning RNA-guided gene drives for the alteration of wild populations. *Elife.* e03401.

17. Selden, S (2005) Transforming better babies into fitter families: archival resources and the history of the American Eugenics Movement, 1908–1930. *American Philosophical Society.* **149**(2):199–225.

18. Gantz, VM, Bier, E (2015) The mutagenic chain reaction: a method for converting heterozygous to homozygous mutations. *Science.* **348**(6233): 442–444. DOI:10.1126/science.aaa5945.

Chapter 8

Cultural Views on Human Genetic Modification

"The only way things will change is if we're smart enough to develop technology that can think us out of this, meaning augmenting ourselves genetically to be smart enough to change..."

— Neill Blomkamp, movie director.[i]

"We are unfashioned creatures, but half made up, if one wiser, better, dearer than ourselves — such a friend ought to be — do not lend his aid to perfectionate our weak and faulty natures."

— Dr. Victor Frankenstein,
character in Shelley's *Frankenstein*

Public perceptions of human modification

Is the creation of GM humans the ticket for humanity to survive an ever-changing, potentially catastrophic future? Or could human genetic modification be more likely to cause disaster or perhaps even the extinction of the human race?

---

[i] http://bit.ly/1GWsBbS

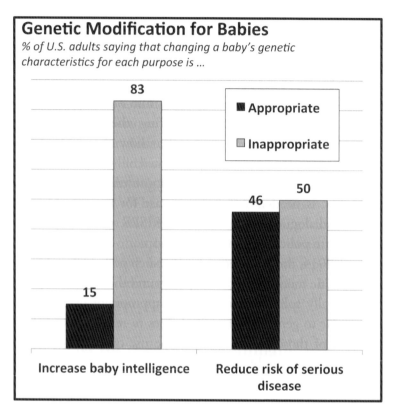

Figure 8.1. Results of a Pew Research Poll on the genetic modification of human babies. US adults were polled in August 2014. Those saying "don't know" are not shown. Image source: "Public and Scientists' Views on Science and Society" Pew Research Center, Washington, DC. Reproduced with permission.

genetic modification of humans[ix] show very little public support. Americans generally are opposed to reproductive sex selection as well.

Global views on human genetic modification

What about the views on human genetic modification in other countries from around the world? Are there some countries where the culture is more clearly supportive of heritable human genetic modification? The

---

[ix] http://www.geneticsandsociety.org/article.php?id=401#igmdata

answer is generally "no." Although recent polling is rather spotty on a global scale, over the past 10–15 years, polling across a variety of countries has indicated fairly widespread opposition to human genetic modification. The pattern of laws around the world also reflects opposition to the intentional use of technology in order to change the human genetic makeup in a heritable manner.

The Center for Genetics and Society has advocated for a moratorium on human genetic modification in the US. In a 2015 policy statement,[x] they pointed out that there currently is no US law on human genetic modification. They also highlighted the fact that many other nations prohibit this sort of work, making it unusual that the US retains a permissive legal attitude toward human genetic modification:

> "The United States should join the dozens of other countries that have passed laws explicitly prohibiting the creation of genetically modified human beings…we also need expanded international agreements, along the lines of the Council of Europe's Convention on Biomedicine and Human Rights[xi] and UNESCO's Universal Declaration on the Human Genome and Human Rights."[xii]

In the current climate, it would be relatively straightforward for a determined research group to try to make a GMO sapiens using private money. Even so, especially if things do not go well, they could face legal consequences. It remains uncertain how it would play out in the US in the absence of clear American federal laws, assuming the team did this in a state that does not explicitly prohibit such work.

Professor Tetsuya Ishii has researched the global legal and regulatory climate on human genetic modification. Dr. Ishii has published a map reflecting the most current regulations across the world (Figure 8.2) [2]. It is important to keep in mind that the regulatory climate can rapidly change. For instance, since the making of this map the UK has now legalized germline genetic modification in the form of three-person IVF

---

[x] http://bit.ly/1cCDvfp
[xi] http://conventions.coe.int/Treaty/en/Treaties/Html/164.htm
[xii] http://bit.ly/1JXT7Yw

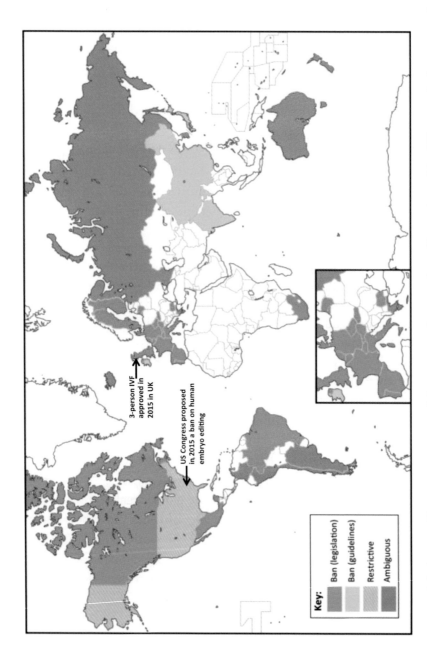

Figure 8.2.   Global map of human genetic modification regulatory sphere illustrating wide differences in laws around the world. Image copyright Dr. Tetsuya Ishii; used with permission.

Figure 8.3. Frankenstein, as portrayed by Boris Karloff in a movie, is the monster that many of us bring to mind when we read or hear the Frankenstein name. Image source: Wikimedia.[xiii]

as discussed in Chapter 4. The climate in China also has some ambiguity and it was there that the first GM human embryos were produced via CRISPR in 2015.

Frankenstein revisited

In Mary Wollstonecraft Shelley's visionary 1818 masterpiece *Frankenstein*, she imagined the non-sexual, DIY-style creation of a human being, who ended up being "the monster." Dr. Victor Frankenstein was the creator. It is worth noting that the monster is now simply called by his creator's name in popular culture and referred to as "Frankenstein" (Figure 8.3).

[xiii] http://bit.ly/1EYS1Hv

The creator is portrayed as having gained powerful knowledge and pioneered new technology, but with disastrous consequences:

> "One man's life or death were but a small price to pay for the acquirement of the knowledge which I sought, for the dominion I should acquire and transmit over the elemental foes of our race."
>
> "So much has been done, exclaimed the soul of Frankenstein — more, far more, will I achieve; treading in the steps already marked, I will pioneer a new way, explore unknown powers, and unfold to the world the deepest mysteries of creation."

Shelley devised a tale that gripped the reader's imagination and brought into play the notion of man as creator. The playing of God by Victor Frankenstein was an act of hubris, but importantly the fact that he did not take care of the monster he created also came into play as a source of darkness.

If people in the real world soon create new types of GM human beings in a laboratory in an asexual manner, an important question is: who would be the new child's parents? Who would care for them, particularly if they turn out to have developmental disabilities, diseases, or deformities? What would be their ultimate fates?

Could GMO sapiens come to be viewed in an intensely negative way, much like new Frankenstein's monsters? Even if not to that extreme, could they be discriminated against? Alternatively, GMO sapiens could be perceived as being clearly superior to regular humans and become a sort of upper class. Essentially, finding the "sweet spot" where GMO sapiens would be viewed as ordinary citizens, who could lead normal, positive lives might prove very difficult or impossible.

The Brothers Huxley

More than a hundred years after *Frankenstein*, we find the cultural world of the early 20th century tackling some of the same issues, but from a perspective in which science was closer to having the power to create humans.

The British brothers Aldous and Julian Huxley were two of the foremost thinkers on human engineering in the first half of the 20th century. Together, their different views embody the dilemmas facing human beings in a technological era in which people may well become equipped to play God with their own creations.

In both fictional and non-fictional writings, the two brothers anticipated many of the challenges we face today. Despite being brothers, their perspectives — as articulated in their writings on future technological innovation and the changing of humanity — could not have been more different.

Aldous penned the famous satirical, dystopian novel *Brave New World* that envisioned a toxic future in which humans use technology to control and warp their own creation with catastrophic effects. Julian had quite a different view. He was a transhumanist and eugenicist, dedicated to the proposition that humans should become better by transforming into a new sort of being. This goal could be most effectively achieved through science and technology, Julian argued.

From a technological perspective, the reproductive methods in Aldous' 1931 novel are remarkably prescient considering it was published more than 40 years before the first successful human IVF procedure by Robert Edwards. Aldous created a fictional universe in *Brave New World* in which the government controls humanity in part through institutional reproduction and human modification.

In the *Brave New World* dystopia, new human beings are produced by the State in facilities such as the "CENTRAL LONDON HATCHERY AND CONDITIONING CENTRE" (named in all caps in the book). This facility also includes a Fertilization Room, where it appears a process similar to IVF occurs on a huge scale. In this future, human creation and modification (in this case non-genetic) in a factory-like setting contribute to a dystopian world pervaded by threads of darkness.

Although today's transhumanists and other advocates of designer babies assert that human modification would not lead to reduced human diversity, in the *Brave New World* universe, the government intentionally creates only a handful of classes of human beings.

In brother Julian's world vision as represented in his writing, which focused on a potential non-fictional future reality, technology could lead to a totally different reality than that described in *Brave New World*, with instead better human beings and a Utopian society. For example, Julian wrote in his essay "Transhumanism":

"I believe in transhumanism: once there are enough people who can truly say that, the human species will be on the threshold of a new kind of existence, as different from ours as ours is from that of Peking man. It will at last be consciously fulfilling its real destiny."

In the polar views of the Huxley brothers' writings, we can see the tension between the power of human modification and reproductive technologies to do both good and bad. These kinds of conflicting views are present in art and literature around the world. It is widely thought that *Brave New World* was at least in part inspired or influenced by Russian writer Yevgeny Zamyatin's dystopian novel *We*, which also imagined a totalitarian world in which reproduction is controlled.

One critical difference between these fictional works and today's reality is that advances in biotechnology make the creation of human clones and GM humans quite possible for society in the coming years. And yet astonishingly, it appears that culturally we are almost entirely unprepared. In addition, the polling discussed earlier indicates the public does not really want to go down that road and take the associated risks.

GATTACA

A more recent fictional work, the movie *GATTACA*, describes a dystopian future specifically based on genetics. In this fictional world, "valids" are a class of people grown from embryos. They are defined as being "better" and are selected by PGD-based genetic testing from groups of embryos. This genetic upper class is considered to have the "right genetic stuff," while "invalids" are wild type humans (as most of us are today) that were born via sex without any laboratory screening and are considered inferior. This is reminiscent of geneticist Lee Silver's vision for the future with a genetic upper class called the GenRich.

In *GATTACA*, we see a eugenic world in which citizens are pressured, including by peers and society, to produce what are perceived to be genetically superior offspring via PGD. While the government does not

Figure 8.4. A movie poster for *GATTACA*, a movie that portrays a genetic dystopian future where people are judged based on their genetics and many humans are made based on embryo choice via PGD. The poster shows the movie's main characters played by Ethan Hawke (top) and Uma Thurman (bottom), who fall in love despite Hawke's character's perceived genetic inferiority. Image © Columbia. Image source Everett Collection/Rex Shutterstock. Reproduced with Permission.

entirely control reproduction, as it does in *Brave New World*, a regimented class society exists in the *GATTACA* universe based on genetic identity and enforced by mandatory routine genetic testing. For example, when main character Vincent (see Figure 8.4, played by Ethan

Hawke opposite his love interest in the movie, played by Uma Thurman), posing as a valid, goes to work, he must fake a blood-based genetic test to enter the workplace. To sneak into work, he uses the DNA of a valid character named Jerome, who was crippled by an accident, instead of his own.

In principle, the same toxic, genetics-based class issues could arise in the real world if we start making GMO sapiens. Even without genetic modifications, PGD could become much more common and influence human society and evolution in major ways. The statement in the movie poster, "There is no gene for the human spirit" resonates with me.

DNA Dreams and reality

In 2013, rumors circulated that China was working toward engineering genius babies.[xiv] The idea went viral. Some parts of the story in the online publication *Vice* had elements of truth to them including that the company BGI, the global genome sequencing giant, is collecting and analyzing DNA samples from some of the "world's smartest people." One of their goals is to find genes and genomic regions that associate with increased intelligence. What happens next if scientists can find a group of gene variants that associate with intelligence? Do they then try to create geniuses?

The online magazine *Slate* took a stab at debunking the *Vice* story.[xv] However, they also pointed out that PGD, for example, is a very real and potentially explosive technology if combined with genetic data on human traits. A few of the experts quoted by *Slate* had some unusual notions, including tying genetic modification to national economic policy:

> "As NYU evolutionary psychologist Geoffrey Miller — a participant
> in the Chinese genome-sequencing study — tells *Vice*, "Even if it only
> boosts the average kid by five IQ points, that's a huge difference in terms
> of economic productivity, the competitiveness of the country, how many

---

[xiv] http://bit.ly/1jESTam
[xv] http://slate.me/1Fd2H8P

patents they get, how their businesses are run, and how innovative their economy is."

In that *GATTACA*-esque view, PGD-based selection for intelligence in China, and perhaps other countries, could become a real economic issue. Some have even suggested that there could be a human genetic modification or genetic "arms race" between countries.[xvi] It is not difficult to imagine military leaders worrying that the armed forces of other countries could become, through genetic modification, staffed by GMO sapiens with super human powers of intelligence, healing, or strength.

Still, the formula for human intelligence is likely to prove elusive. It could require such a complex genetic recipe as to be nearly impossible to successfully create intentionally. The same *Slate* article cited earlier weighs in on this as well:

"Hank Greely, director of Stanford's Center for Law and the Biosciences: "I think it's pretty clear that intelligence — if it even exists as an entity, which remains controversial among psychologists — involves a boatload of genes and genetic combinations, all of them substantially mediated through the environment. The chances that genetic selection is going to lead to really substantial increases in human intelligence in your lifetime are low."

Greely believes that successfully selecting for human intelligence is a long shot. Geneticist Lee Silver of Princeton, who was also quoted in the piece, was largely in agreement with Hank's view, but is not a fan of leaving human reproduction up to Mother Nature:

"In my opinion, even a partially informed choice is always better than chance," he says. "Those who reject this point of view often don't think of the natural process as chance, but rather as God or Mother Nature doing her work. But as I said to Stephen Colbert on his show, 'Mother Nature is a nasty bitch.'"

[xvi] http://www.geneticsandsociety.org/article.php?id=8500

Clearly, PGD as a technology has a role in preventing genetic disease, but its use for trait selection is poised to become an intensely controversial issue in the next few years. *GATTACA*, sometimes viewed as portraying a cliche dystopia, might not be so far off from the actual reality that we could see before the decade is out.

Getting back to the Chinese sequencing company BGI, there are signs that it wants to do more than just obtain genomic sequence information on intelligence and other human traits. In Dutch filmmaker Bregtje van der Haak's amazing documentary film *DNA Dreams*, a different cultural aspect to BGI is revealed. As we watch and listen in during her film, BGI scientists chat about what they want to get out of genomic technology:[xvii]

> "This isn't even positive eugenics that we're talking about, we're not encouraging smart people to have kids, we're encouraging everyone who has kids to have the best kids they possibly could have."

That kind of goal sounds similar to eugenics and again this above quote from one of the "key BGI players"[xviii] raises the question of what the "best kids" as determined by genetics would be? In the case of the BGI Cognitive Genomics Group, it is focusing on the "best" children from an IQ test-based perspective.

Theoretical physicist Stephen Hsu of Michigan State University, is more blunt than others interviewed in the film about the jump to making "super" humans via genomics and genetics:

> "The best humans have not been produced yet, the smartest humans, the longest lived humans…If you want to produce smart humans, nice humans, honorable humans, caring humans, whatever it is, those are traits that are related to the presence or absence of certain genes and we'll have much finer control over the types of people that are born in the future through this…which switches do you turn on or off to do that? …We do it with cows, we have super cows and super chickens…We've pushed

those animals in directions we want to push them, but we haven't really pushed ourselves, and I think people will push themselves."

When it comes to genomics and human traits, "best" is in the eye of the beholder. Clearly some scientists in countries around the world including here in the US, in China, and elsewhere want to make efforts at making better and the "best" humans. Or even super humans.

DNA Dreams not only focuses on genomics and the scientists at BGI (who, by the way, are a brilliant, creative, energetic bunch), but also on large-scale animal cloning work going on at BGI. I wonder if some of the human genomics work and the cloning efforts could come together ultimately some years in the future to lead to creation of GMO sapiens.

Towards the end of *DNA Dreams,* one of the BGI scientists, after pointing out the complexity of genetic control of human intelligence nonetheless says, "But people ought to be free to manipulate their children's IQ. It's their own choice."

In China, the range of reactions to the first human embryo editing science paper published there was mixed. However, overall it pointed generally toward a scientific and bioethical philosophy quite different in some ways than that of the US and Europe.[xix] In an interview with the *New York Times* for an article on that first embryo editing paper, Deng Rui, a medical ethicist at Shanxi Medical University, pointed to some key differences:

> "Confucian thinking says that someone becomes a person after they are born. That is different from the United States or other countries with a Christian influence, where because of religion they may feel research on embryos is not O.K."

China's National Medical Ethics Committee approved the human embryo editing work reported in that published paper at least in part because it was considered "not for reproductive purposes," but that line could be crossed in China, the US and elsewhere in coming years.

[xix] http://nyti.ms/1ekpe7s

The *New York Times* piece indicates that more ethically contentious types of genetics work in China are likely already ongoing, quoting several unnamed scientists that, "More unpleasant scientific surprises are looming." Rao Yi, a professor of biology at Peking University, believes that more human embryo editing will be reported from China:

> "Right now, human gene editing is the main thing," Mr. Yi said. Geneticists in China "don't want to be guided by Western people."
>
> The mind-set among Chinese researchers, according to Mr. Yi: "'We're going to do it, then see what's wrong, then fix it. But the conceptual discussion may be missing.'"

Even as the cultural differences are fascinating, as human modification technology and research expand, we can expect more debate and even clashes.

Orphan Black

The ongoing BBC America TV series *Orphan Black* brilliantly portrays a series of female and male human clones. They are unwitting parts of an experiment and face danger, including the threat of being used or even killed by religious extremists who view them as abominations. The corporation that is doing the cloning in the world of *Orphan Black*, the Dyad Institute, has some link to the government. However, it is making clones secretly and does not want the cloning to become public.

The clones of *Orphan Black* also have genetic modifications in the form of "barcodes" that distinguish them from each other, important since numerous genetically identical clones of each type are made in that world. Barcodes are a real thing in genomics research today. When we sequence different samples in the lab, we can identify them even if mixed together based on the unique barcodes we build into the sequences. During natural DNA analysis studies, geneticists can identify the different species contributing to a mixed DNA sample by way of specific, naturally-occuring genetic elements called "barcodes" as well. Genetically introduced barcodes could be used for identification of

GMO sapiens and in principle could cause problems via changes in DNA sequence.

A surprising twist to the barcodes in *Orphan Black* is that the clones are viewed as property because the technology has been patented. The tags also identify them as commercial property. Two clones, Cosima and Sarah, discuss this odd reality:

> *"**Cosima Niehaus**: Any freedom they promise is bullshit. They're liars. That synthetic sequence, the barcode I told you about? It's a patent.
> **Sarah Manning**: A patent?
> **Cosima Niehaus**: We're property. Our bodies, our biology, everything we are, everything we become, belongs to them. Sarah, they could claim Kira."*

Even the clone Sarah's daughter Kira might be viewed as property as well, as a "derivative" of a clone. This fictional portrayal in *Orphan Black* raises a potential issue in a future real world where humans are being genetically modified: Would GMO sapiens come to be viewed as intellectual property? Is it possible that someone could claim ownership of GMO sapiens?

In the US, laws prohibiting slavery would rightly block this.[xx] The recent US Supreme Court ruling against the biotech company Myriad, which had patented the *BRCA1* gene, means that genes also cannot be patented in the US. However, the unique DNA sequences in GMO sapiens created by the genetic modification process likely could be patented. In that sense, that particular part of a GMO sapiens (and of its future descendants) could be classified as property.

A difficult question then arises. What happens if GMO sapiens seek to reproduce and in so doing the patented GM sequence is reproduced as well in their GM children? Could that attempt to replicate the patented DNA even indirectly via production of a child be viewed as a form of piracy?

Watching *Orphan Black* with all its female clones so believably portrayed by the one actress Tatiana Maslany can be a bit mind-bending for the viewer.

[xx] http://io9.com/the-real-life-science-behind-orphan-black-1694765437

Maslany does an amazing job at what must be far more difficult: portraying the entire series of female clones who all have different personalities, accents, and characteristics. The diversity of traits and perspectives of each clone also brings home a powerful message from the creators of this show: genetics is not destiny. For instance, in addition to heterosexual clones, there is a transgender clone and a lesbian clone. Over the years in the real world, some have claimed controversially that there could be a strong genetic basis to sexual orientation or that a "gay gene" could exist.[xxi] I am skeptical about that. Genetic determination more generally is a simplistic notion at odds with science.

The diversity of the clone phenotypes on this fictional show overall also raises the real point that attempts at human genetic modification or cloning are likely to have unpredictable results due to powerful environmental influences and the role of chance.

An artist's view of human cloning

The words and imagery of human cloning and modification are very pervasive in art and literature. One of the more powerful artistic renderings of cloning that I have seen recently comes from artist Daisuke Takakura (高倉大輔) (see Figure 8.5).

Takakura's work strongly brings home what a world with clones might feel like.

Genetic discrimination or celebrity

In the past, Americans have already been concerned about possible discrimination based on a person's genetics. This discrimination could take many forms, such as higher insurance rates or an inability to be employed. In response to such concerns, the US passed the Genetic Information Nondiscrimination ACT (GINA)[xxii] in 2008, signed into law by then President George W. Bush.

[xxi] http://bit.ly/1vpiYma

[xxii] http://1.usa.gov/1JXTrXi

Figure 8.5. A graphic depiction of human clones by artist Daisuke Takakura (高倉 大輔), Image source and copyright: Daisuke Takakura. Used with permission.

Oddly, the sole dissenting vote against GINA in the US Congress was Congressman and physician Ron Paul. Business groups, including the US Chamber of Congress, opposed the bill as well, feeling it might be too restrictive. Insurance companies were also generally not happy with GINA and have indicated a desire to use genetic information as the basis for determining rates for consumers.

GINA was a positive step in providing assurances that people would be free from some forms of discrimination based on their genetics. For example, GINA prohibits health insurance companies from using genetic information to determine coverage eligibility or to assign higher premiums based on a customer's genetics. Employers are also prohibited from using genetic information.

At the same time, there are some notable gaps in GINA, including the fact that life, long-term care, and disability insurance companies are basically exempt from these rules. As a result, they legally can determine premiums or even deny coverage to a consumer based on their genetic state. For this reason, genetic discrimination is probably already occurring in the US in certain forms related to insurance.

These exceptions as well as other potential forms of discrimination based on genetics are concerning. At the same time as people are afforded rapidly increasing opportunities to obtain genetic information about themselves and their family members (either via their own physicians or by companies such as 23andMe), it remains unclear how that information might be used or misused in the future by others.

In this context, if designer babies were to enter into this system, how would they be treated? On the one hand, they might be discriminated against because of their unique status as having been genetically modified. However, if it comes to be believed in the wider culture that GM people are healthier or somehow otherwise "better" than average they could come to constitute what might be called a "genetic upper class," as cited earlier in this chapter in the context of fiction.

Take the case of a designer baby girl who in the first eighteen years of her life never gets sick or injured, and who never causes any trouble. If the same child excels in the school, is attractive, and a model citizen, she might quickly become a "poster child" for promoting genetic modification. I believe that she, and other GMO sapiens like her, would likely become celebrities, driving a pop culture trend that is supportive of human genetic modification. In a sense, this could create a sort of cultural gene drive for human genetic modification.

Some parents with disabilities have specifically selected embryos in order to have children who have genetic states that confer the same condition on their children.[xxiii] This choice resonates with concerns about pushes for designer babies and also about ableism, a form of discrimination against people with disabilities and in favor of the able-bodied. A new form of ableism empowered by genetic modification technology, a modern eugenics, could leave those of us who are unmodified as a genetic lower class.

[xxiii] http://nyti.ms/1J2tCp6

Gender issues in human modification

You may have noticed in reading this book that there have been notable, concerning gender dynamics at play in the arena of assisted reproduction, human genetic modification, and cloning over the years. The history in this area of biomedicine is skewed toward stereotypical gender roles. We saw and continue to see mainly male scientists and physicians conducting experiments that disproportionately involve women as subjects or that rely upon female reproductive tissues or cells.

In the community of stakeholders who have weighed in on human genetic modification, Duke Law Professor Nita Farahany is one of the few women out there supportive of the idea for three-person IVF and possibly under certain conditions in the future of germline genetic modification of specific genes:

> "People should be able to make a vast array of choices," she said. "It's unlikely to lead to state-sponsored eugenics, because private choices by private individuals are going to vary."[xxiv]

I am less certain that eugenics is unlikely to result from human genetic modification.

Why more broadly is there an apparent divide between female and male perspectives on human genetic modification? Could it be a representation of long-held stereotypical roles in biomedical sciences?

Regardless of the cause, this dynamic is problematic and raises important ethical issues as these related areas of experimental human research continue. For example, if *in vitro* research on CRISPR-based genetic modification of human embryos rapidly picks up pace, the resulting need for large numbers of human eggs and embryos could be very ethically problematic.

Even specifically only in the area of three-person IVF/mitochondrial transfer, a significant number of human eggs will be required from donors just in the UK. As Mitalipov and others are likely to do similar work in China, there will be increasing demand for human eggs there as well.

[xxiv] http://theatln.tc/TMdj7K

The net result of these trends moving forward could be a sense of commodification of women and their eggs related to human modification. If research goes further to include attempts at actual production of designer babies, we can expect that in some cases the biological mother may not want to or be able to produce her own viable eggs.

Future mothers of designer babies may also in some cases not be able to carry the babies to term themselves or choose not to do so. In those kinds of cases, many surrogate mothers will be needed to produce the designer babies, raising additional ethical concerns.

Parents-to-be, if equipped with powerful genetic modification technology, may also make some unfortunate decisions. They may choose gender-stereotypic traits for their female and male offspring, literally changing a stereotype into a reality for these children. We could end up seeing children produced who have wildly exaggerated stereotypical male or female traits. Consider for comparison the extreme lengths to which some people go with cosmetic procedures to change their appearance to look a certain way that fits with their preconceived notions of gender.

The nexus of embryonic and fetal screening procedures with issues of gender, reproductive rights, and autonomy also raises concerns. A troubling element in this area is pointed out by Alexandra Minna Stern in her 2005 book *Eugenic Nation: Faults and Frontiers of Better Breeding in Modern America* [1]:

> "Finally, there is another serious medical and moral paradox that shows little hope of resolution in the near future. Over the past three decades, genetic and reproductive technologies have rapidly expanded, providing American women with an increasing amount of information about potentially inheritable diseases and the existence of chromosomal abnormalities. Yet at the same time, the main procedure that a woman can choose to terminate a pregnancy, namely abortion, has been gradually recriminalized on both the state and federal levels."

If genetic testing continues to increase in popularity and the availability of safe abortions declines further, this troubling situation for women could grow far worse.

Although PGD screening occurs prior to pregnancy, there is some concern that efforts at human germline genetic modification could well lead to more conflicts including during pregnancy as incorrectly gene edited or otherwise "problematic" fetuses are identified.

There is value in being aware of these gender issues. Furthermore, discussing possible solutions for avoiding commodification of women as well as other gender-related problems linked to these technologies will be vital as they could have tremendous negative consequences.

Letting your GMO imagination run wild

If you set aside any moral, ethical, or other concern about human genetic modification, and let your imagination run wild just for a moment, all kinds of extraordinary possibilities could arise from human genetic modification. Totally new types of people could be made in such a fearless new world. It could be both wondrous and disastrous. I am inclined to believe it would be more disastrous, but it is thought-provoking and even fun to think about the possibilities from a strictly hypothetical perspective.

GMO superheroes and supervillains

The notion of god-like creatures endowed with super powers, as well as the idea of combining humans with other creatures, have been popular in art and mythology for thousands of years. Our forbearers conceived of all kinds of human-animal and animal-animal combinations such as mermaids, centaurs, Pegasus and griffins. In principle, genetic modification technologies and reproductive technologies could allow for attempts at the creation of real chimeras, superheroes, and villains in the future.

Chimeric embryos involving human and animal cells have already been produced in some research laboratories, but for example in the US this chimera research is now strictly regulated and such hybrid embryos are only allowed to develop for a limited period of time. Technological advancements in embryology over the last few decades have brought the possibility of creating such chimeras closer to reality. The notion of combining humans and animals is very controversial, but the idea could prove seductive. For, instance a team led by Professor Steve Goldman of the University of Rochester has found that chimeric mice with some human brain cells are significantly smarter than ordinary mice [3].

There is a whole spectrum of human superheroes with extraordinary powers in art and some are chimeras of a sort while others are not. GM technology makes the idea of human superheroes feel a bit closer to reality. Science fiction and fantasy have taken their turns at such inspired flights of imagination, endowing humans in art with attributes of other chimeric-like creatures, such as SpiderMan.

Want to make a person like the superhero Aquaman? Make a GMO sapiens with gene insertions that developmentally drive gill formation. Would you get an amphibian human? In principle it could work, but the reality could be a disastrous mess.

What about a person with animal vision, such as super acuity or night vision via genetic modification technology? How about a human with incredible animal speed? Usain Bolt would not stand a chance against a person with the muscle physiology of a cheetah, for example.

Recently, a reader of my blog, Brian Sanderson, wrote a startling comment, discussing the idea of GMO soldiers:[xxv]

> "It seems to me that it is just a matter of time before technologies get to the point where genetically modified humans (GMH) can be 'created'. Do we really want to go there?
>
> It certainly opens up some interesting new possibilities for warfare. Imagine the creation of a strain of 'roundup ready humans' that are resistant to some secretly created pathogen that is fatal to 'natural' humans…"

That is a disturbing thought. The military could be interested in producing GM human technology as a matter of perceived national security, leaving those of us in the civilian world out of the loop. Could GM technology already be part of military discussions looking to the future?

[xxv] http://www.ipscell.com/2015/04/roundup-ready-humans/

Figure 8.6. An illustration depicting the propaganda-like intensity of the debate over Monsanto and GMOs. Copyright, Benjamin Karis-Nix. Reproduced with permission. Source BenKN.com

In some sense there has already been a world GMO war ongoing, but this one surrounding GM plants and foods. There are many people with intense feelings on both sides of the GMO debate. For instance, this illustration by artist Benjamin Karis-Nix (Figure 8.6) is meant to reflect the combative and even propaganda-like intensity invoked against the GMO company, Monsanto. Those on the pro-GMO side of this debate also often voice intense feelings in defense of genetic engineering. In this book, our focus is instead specifically on the growing debate over human genetic modification.

How will culture view real GMO sapiens?

If new, real GM humans come onto the scene, much of how they will be treated culturally will depend on how and why they are modified. Were they created simply to avoid a genetic disease? If so, then I believe they will be treated respectfully and much as with IVF-created people, they will be considered normal members of society, after a period of cultural adjustment. This will to a large extent depend on how they turned out as well.

Moving one step further, if people are genetically enhanced with traits that nonetheless are still within the spectrum of the human norm (e.g. creating a new person with 20/20 vision to replace nearsightedness rather than giving them eagle vision, a type of example given by Nita Farahany)[xxvi] that may still prove controversial but over time would also come to be considered largely a normal practice.

Even more controversial would be creation of GMO sapiens strictly for the purpose of more extreme enhancement. I predict there may well be cultural and perhaps even legal conflicts that would arise in this last scenario. The chance for personal human tragedies also would be greatly increased.

As cutting edge technologies increase the possibility of real human cloning and genetic modification, art and cultural views are likely to continue to provide insights into the possible impact on humanity and for better or worse, culture will adjust. Today's cautious views of GM humans could come to be viewed as quaintly out of date or even anti-science in the future. For centuries humanity has been grappling in different ways with taking control of human creation. The difference today is that the technology that makes it possible now exists, is easy to use, and widely available. The decisions that we make today on this front will profoundly alter our species and the future of our world.

References

1. Stern, A (2005) *Eugenic Nation: Faults and Frontiers of Better Breeding in Modern America*. University of California Press.
2. Ishii, T (2015) Germline genome-editing research and its socioethical implications. *Trends in Molecular Medicine.* **21**(8):473–481.
3. Windrem, MS, Schanz, SJ, Marrow, C, Munir, J, Chandler-Militello, D, Wang, S, Goldman, SA (2014) A competitive advantage by neonatally engrafted human glial progenitors yields mice whose brains are chimeric for human glia. *J Neurosci.* **34**(48):16153–16161.

[xxvi] http://www.ipscell.com/2015/07/farahany/

Chapter 9

GMO Sapiens Today and Tomorrow

"The new technology enables parents to make choices about their children just as they might with Ritalin or cleft palate surgery to 'improve' behavior or appearance."

— George Church, Harvard Professor[i]

"We are humans, not transgenic rats… We believe there is a fundamental ethical issue in crossing the boundary to modifying the human germ line."

— Edward Lanphier, President of Sangamo[ii]

Creation of the first gene-edited human embryos

In 2014 in Dr. Junjiu Huang's laboratory at Sun-Yat Sen University in Guangdong, China a scientist sat hunched over a special microscope staring at one-cell human embryos in a dish. They would appear like tiny translucent spheres glowing in the light of the microscope. The scientist guided a very thin, sharp needle to pierce the first of a series of such human embryos.

---

[i] http://www.ipscell.com/2015/03/georgechurchinterview/
[ii] http://bit.ly/1LM7GQL

An almost immeasurably small amount of liquid, but carrying within it very power technology, flowed through that needle into the embryo. CRISPR-Cas9 gene editing molecules went from the outside world into that human embryo. The needle came out and CRISPR-Cas9, that Swiss Army knife of genetic tools, stayed inside. It rapidly found its way to the embryo's DNA where it began cutting and editing. Other human embryos were injected to become GM too. Note that the ooplasm work discussed earlier in the book also created some GM human embryos and GM humans more than a decade ago, but that was done by using a somewhat blunt force method by injecting into eggs some random amount of cytoplasm containing mitochondrial DNA from a donor egg (refer back to Chapter 4).

Over the years, specific biologists and doctors have at times taken big risks and pierced other barriers to conduct controversial human experiments. They often broke convention and their work frequently sparked debates as well as concerns over ethics. Not all such risk takers were truly innovative or successful. I have written about some of the most influential individuals and their historic work in earlier chapters.

In the embryonic stages of writing this book, rumors had been flying that researchers in China, and possibly elsewhere, were trying to publish work that reported the gene-editing and genetic modification of human embryos. We know now that the rumors were right.[iii]

On April 18, 2015, Dr. Huang's team published their article on the creation of GM human embryos in the journal *Protein & Cell*. The human embryos that had been edited were obtained from a fertility clinic. It is notable that this work was done at Sun Yat-Sen University because some of the earliest human experiments that would have created GM humans (via SCNT used for trying to prevent mitochondrial disorders) also occurred at the same institution with terrible results (see Chapter 4). With the new human embryo editing work, Sun Yat-Sen University continued a tradition of controversial human embryo research. It is possible that there have been additional attempts by other teams elsewhere to make GM human embryos that failed and were never published, leaving the world unaware.

[iii] http://www.ipscell.com/2015/04/gmhumanembryos/

According to an article in *Nature News* by science writers David Cyranoski and Sara Reardon,[iv] the Huang lab human embryo editing paper published in 2015 had been submitted earlier and rejected at top journals, such as *Nature* and *Science* due at least in part to "ethical issues." I had heard the same thing. However, as I am writing this book now no further details are available on that.

Nature News quoted pioneering scientist George Daley on this development:

> "I believe this is the first report of CRISPR/Cas9 applied to human pre-implantation embryos and as such the study is a landmark, as well as a cautionary tale," says George Daley, a stem-cell biologist at Harvard Medical School in Boston. "Their study should be a stern warning to any practitioner who thinks the technology is ready for testing to eradicate disease genes."

The GM human embryo paper was entitled "CRISPR/Cas9-mediated gene editing in human tripronuclear zygotes."[v] These human embryos were not viable, meaning that they could never make a human being. The authors apparently sought this relatively more ethically acceptable way to go for this first published attempt at editing the genome of human embryos. This particular type of abnormal human embryo cannot develop fully into a person because two sperm fertilized it back when it was an egg. In technical terms, they are called "3PN embryos."

The specific gene-editing goal of this work was to correct a mutation in the β-globin gene (a gene also known as *HBB*) that causes the human disease beta thalassemia. The team reported that this CRISPR gene editing in the human embryos did not go well. It mostly did not work to fix the beta thalassemia mutation and even when the proper edits were introduced, the results were mosaic such that only some cells had the edits, while others in the embryo did not. To make matters worse technically speaking, there were high levels of off-target activity: "These data demonstrate that CRISPR/Cas9 has notable off-target effects in human 3PN embryos." In

---

[iv] http://bit.ly/1DCtGn5
[v] http://link.springer.com/article/10.1007%2Fs13238-015-0153-5

other words, the CRISPR-Cas9 gene editing machinery often made edits in the wrong place. Sometimes it made the wrong kinds of edits too in the right place. You can refer back to Chapter 6 for my discussion of such errors.

Overall, this attempt at embryo editing was largely unsuccessful. While critics point towards the fact that the team used already somewhat outdated CRISPR-Cas9 tools and approaches, this outcome nonetheless suggests that the level of precision required for clinical use of CRISPR-Cas9 in humans may be difficult or even impossible to achieve.

Was this work ethical? Would an institutional review board in another country, such as the US, have given the green light to making these GM human embryos? While so far nothing clearly unethical is apparent in this published work, it still raised concerns and even outrage in some cases.

The *New York Times* published a piece on this work and noted how it relates to the broader climate of bioethics in China.[vi] This article quoted a number of scientists, regulators, and ethicists with diverse perspectives on the human embryo genetic modification work and on the paper reporting it including this view:

> "Disturbed by the recent study, Rao Yi, a professor of biology and director of the four-year-old Center of Life Sciences at Peking University, run jointly with Tsinghua University, warned that scientific research in China urgently needed more effective ethical oversight."

In principle, I could support some kinds of *in vitro* work on gene modification in human germ cells and even in early embryos (see my ABCD plan[vii] described later in this chapter). However, the type of work published by the Huang group and the outcomes reported there make me very uncomfortable from an ethical perspective. I am not convinced that this human embryo experiment was even necessary.

As of the publication of this book, several other groups in China (and perhaps elsewhere) are reportedly conducting similar research. Rumor has it that at least one laboratory doing this is using normal or near-normal,

---

vi http://nyti.ms/1ekpe7s
vii http://www.ipscell.com/2015/03/abcd-plan/

viable human embryos that only have specific, discrete, and disease-associated mutations to be targeted for genetic modification.

Even before this new paper came out reporting the first GM human embryos, several groups of scientists proposed moratoriums on various kinds of human editing ranging from banning all work including even that limited just to the laboratory to a recommendation to freeze any clinical applications.

"Don't edit the human germline"

Before the Chinese paper was published, on March 5, 2015 Technology Review broke the story in a piece entitled "Engineering the Perfect Baby" that CRISPR experiments both in the US on human germ cells and potentially in China on the human embryos was ongoing.[viii]

A week later came the first formal sign of scientific concern about possible imminent human embryo editing in the form of a commentary piece published on March 12, 2015 in the journal *Nature*, entitled bluntly enough, "Don't edit the human germline" [1].

It was not included there in the title, but one could well imagine an exclamation point after those words. The article's authors very strongly articulated a case against allowing any genetic modification of human embryos, even if limited just to the laboratory.

One notable element of this "Don't edit" article calling for a prohibition on human germline gene editing was that the authors included prominent gene-editing experts.

They articulated their reasoning against human gene editing in this way:

"In our view, genome editing in human embryos using current technologies could have unpredictable effects on future generations. This makes it dangerous and ethically unacceptable. Such research could be exploited for non-therapeutic modifications... At this early stage, scientists should agree not to modify the DNA of human reproductive cells."

viii http://www.technology review.com/featuredstory/535661/engineering-the-perfect-baby/

These scientists are concerned both about the potential direct consequences of human genetic modification in embryos and about the possible public outcry against such work should it proceed. They also pointed out that the effects of gene editing in humans, if GM humans were created, could not be known until after such babies were born. For example, problems might not surface until later in the lives of a genetically modified individual or when they attempt to have children themselves.

Even earlier than the Lanphier article, bioethicist Tetsuya Ishii articulated concerns about possible heritable human genetic modification in 2014 [6], and Professor Anthony Perry's group, which conducted genetic modification of one-cell mouse embryos using CRISPR, also raised the issue of potential human modification in a research article [7] that same year as well.

"Prudent path forward"

Only a few weeks after the Lanphier article was published, a different group of respected researchers published another commentary. This one had a relatively more moderate tone on human germline modification.

Nobel Laureate David Baltimore and co-authors, including CRISPR pioneer Jennifer Doudna, Stanford Professor Hank Greely, and geneticist George Church charted a potential path forward for human genomic engineering involving germline modification in their piece in the journal *Science* [2]. I suggest also reading the accompanying *Science* article by Gretchen Vogel, "Embryo engineering alarm" [3].

The Baltimore piece was entitled "A prudent path forward for genomic engineering and germline gene modification." The key word there is "forward" suggesting the authors do not want an absolute moratorium on human gene editing. The article called for further discussion and assessment of key potential benefits and risks to moving forward with this technology. The piece reflects the conclusions of an earlier meeting in 2015 that the authors of the article had held in Napa, California on this issue.

The summary statement of the Baltimore piece is as follows: "A framework for open discourse on the use of CRISPR-Cas9 technology to manipulate the human genome is urgently needed." This statement is mostly in line with the policy statement released by the International Society for Stem Cell Research (ISSCR) around the same time,[ix] but is perhaps not quite as strong as the ISSCR statement. It does not, for example, call explicitly for a moratorium on clinical use of gene-editing, as the ISSCR does; it instead "strongly discourages" it.

In addition, this Baltimore piece definitely conveyed more optimism and a relatively positive vision that CRISPR-Cas9-mediated human germline editing might have safe, effective, and ethical clinical applications in the future. Even so, they remain appropriately cautious:

> "At present, the potential safety and efficacy issues arising from the use of this technology must be thoroughly investigated and understood before any attempts at human engineering are sanctioned, if ever, for clinical testing."

Two of the authors, Baltimore and Paul Berg, were also present at the 1971 Asilomar conference on DNA modification, which I discuss earlier in the book in Chapter 2. We should consider the potential areas of disagreement amongst this list of scientists and bioethicists, even though they clearly came to a consensus on some key issues in their one collective paper. For example, in Vogel's piece, Church is quoted as having a more relaxed perspective:

> "Those uncertainties, together with existing regulations, are suffi-cient to prevent responsible scientists from attempting any genetically altered babies," says George Church, a molecular geneticist at Harvard Medical School in Boston. Although he signed the *Science* commentary, he says the discussion "strikes me as a bit exaggerated." He maintains that "a de facto moratorium is in place for all technologies until they're proven safe."

I am less trusting on this issue both in terms of all capable scientists being deterred and that we have an existing operational moratorium.

[ix] http://www.ipscell.com/2015/03/isscrgermediting/

Figure 9.1.   CRISPR-Cas9 innovator, Jennifer Doudna, is shown in her laboratory. Image source: US Department of Energy.

My sense is that at present there are no clear societal or legal obstacles to making GM humans.

Doudna and some of the others also appear far more concerned and cautious than Church. I interviewed Doudna (pictured in Figure 9.1) to learn more about the Napa meeting and her perspectives on human germline modification. I have included excerpts of the blog interview on the following pages.

Knoepfler: What specifically sparked the Napa meeting? Did you help to start the ball rolling?

Doudna: The Napa meeting was organized by myself and my colleagues at the Innovative Genomics Initiative. We had ethical concerns regarding potential applications of genome editing because CRISPR-Cas9 is widely adapted and so simple. We felt it was important to convene a meeting of stakeholders.

Knoepfler: At that time did the organizers know about the rumors of papers potentially in review that might report genetic modification of human embryos?

Doudna: I wasn't aware of that work coming up, the possible papers in press on embryo editing. We knew, however, that that type of work was possible. We hoped that before it would appear in a publication we'd be able to pull together a consensus point of view through this meeting. We didn't go in with a preconceived notion, but we felt that scientists should be discussing this, and it should be open and transparent.

Knoepfler: How did the meeting go? Were there some areas of disagreement?

Doudna: It actually went fairly smoothly. There was definitely very animated discussion. This is a topic that people can feel emotion about. It is pretty profound if you talk about clinical applications that could change human evolution. There were different points of view, but not hugely different. I didn't hear anybody at either extreme saying things like "We should edit people tomorrow!" or "We have to get rid of this technology." It was more focused on questions such as "What kind of safety or regulatory matters should be discussed?" It was only a one-day meeting so there wasn't a lot of time to get into other issues such as gene editing triggering a biological chain reaction where a dominant change could spread through a whole population.

Knoepfler: Will there be another Asilomar meeting like the one in the 1970s?

Doudna: That's the goal. We would like to convene a larger meeting. We want broad, international representation. That meeting will probably happen later this year. A number of groups have expressed interest. The aim would be to have representatives of the various stakeholders. Some of the top scientific advisory groups. Governmental groups. Funding agencies.

Knoepfler: George Church seems to have a more open view on application of this technology.

Doudna: Any group of people will have a diversity of opinions. It's the kind of topic that each of us comes to with our own set of beliefs and level of comfort with making changes to the DNA of an organism. That's one of the reasons to get together. I found that Napa meeting to be extremely interesting and stimulating. George was not there in person, but he had a lot of input. I certainly learned a lot.

Knoepfler: Should there be a "pause" on clinical applications? How would that work? I don't think I saw the word "pause" in the *Science* piece.

Doudna: We decided not to use the word "moratorium" because some people view that as policing. How do you police it? That may be hard or impossible. Rather, we were suggesting that the community decide together about the technology. There's this incredible potential to help, and yet also risks. It's a great technology and we should be excited about it. It is incredibly enabling. It enables a lot of new biology that would have been otherwise difficult or impossible. At the same time, because it is powerful and straightforward, it means it also enables clinical and other applications that might be problematic. We, meaning the scientific community, have to proceed in a way that is considered.

We can't dictate what is happening in every part of the world. It would be presumptuous to ask for a moratorium worldwide. To be more realistic, we strongly encourage scientists to not use it for this clinical kind of purpose at this time. Let's make sure we are doing the appropriate research before employing it in ways that could be harmful (or helpful for that matter). The recommendation is for a clinical pause. Research should proceed. Then this provides data to evaluate clinical potential and risks.

Knoepfler: Can you imagine a future point at which you'd support the use of gene editing in humans in a heritable manner? If so, how do we get to that point from where things are today? What do we need to learn first?

Doudna: We need to learn how efficiently it works. What's the best way to deliver it safely and efficiently? Not only efficiency, but also what are the off-target levels? How do we minimize that? What would be a safe level if any of off-targets? I'd like to see basic research like what happens to the DNA in germ cells or pre-germ cells when a double-stranded break occurs? What is the repair process like in those specific cells? Those answers would be interesting from a basic science perspective as well as informing future potential clinical applications.

I feel uncomfortable imagining widespread gene editing use in humans now, but it is possible that there are going to be certain types of very specific applications that could be envisioned as beneficial in the future. I won't be able to make a decision of the wisdom of such an approach until we have more data. What are the real risks? There is always a risk-versus-reward kind of consideration. In which cases is the risk worth the payoff?

Knoepfler: Can you comment on the recent *Nature* piece by Lanphier, *et al.* from ARM and Sangamo? Did you read that they are opposed to any germline editing, even *in vitro*? What do you think of their view?

Doudna: They would not allow research. That's one point of view. The challenge I see with that is first of all I don't see how you would stop it. Secondly, even if everyone agrees with that, how do we move forward? If we don't do the research, how do we make informed decisions on the potential for clinical use? If it is possible to permanently fix a genetic mutation that is responsible for a horrible disease, maybe there would be an argument; if we have the technology to fix it, maybe that's a good thing to do. How do we know if we don't do the research? To say that we are not going to do any research, it blocks the ability to inform.

Knoepfler: What do we do if someone goes rogue?

Doudna: That's one of the purposes of these meetings: to get out in front of that. I can't guarantee that that might not happen. I can work to form a coalition to say, "here's our considered view of the technology and here's what we see as a prudent way to move forward with this." That's really the best that we can do. There's no way to unlearn what is learned. We can't put this technology to bed. If a person has basic knowledge of molecular biology they can do it. It's not realistic to think we can block it…We want to put out there the information that people would need to make an informed decision, to encourage appropriate research and discourage forging ahead with clinical applications that could be dangerous or raise ethical issues.

Knoepfler: Do you have any comments on the patent situation?

Doudna: I think any time there's a new technology there are going to be multiple claims around it. Especially something that's very broadly enabling. That's something that's probably par for the course. What I would love to see as a scientist is to see people helped with this technology and to see society helped. I would like to see this employed to do that. I would hope that the IP situation doesn't impede that. I'm just speaking as a scientist or person. Not as a lawyer or a UC professor.

This interview with Doudna provided some key insights into how the CRISPR-Cas9 technology opens new, exciting and admittedly complicated doors. At the same time, it illustrates just how difficult possible human applications of gene editing technology are likely to be both technically and from a policy standpoint if to be done in a responsible manner. In addition there is a sense of uncertainty about the path ahead including how the technology will evolve, patent issues, and how best to balance regulation and freedom of science.

Hank Greely, was another of the authors of the Baltimore, *et al.* *Science* article [2] along with Doudna. Sometime after the *Science* article came out, as cited in the previous chapter, Greely blogged[x] about the Napa meeting that catalyzed the writing of that article. He articulated why he thought human germline modification was not going to be a "big issue" with the foremost reason being that it would be too unsafe for people to try:

> "First, the safety issues are enormous. That's not to say anything bad about CRISPR-Cas9 or other genome editing techniques, but the stakes are enormous... a human baby... It is not just a matter of perfecting CRISPR-Cas9's targeting ability or any other particular parameter; we'd want to see, in advance, whether other things, previously unanticipated, about the process were likely to cause problems in making babies. You'd have to be criminally reckless, or insane, to try to make a baby this way... If the moral risk isn't enough of a deterrent, the potential legal liability should be."

Ironically, one of the top reasons I believe that human germline modification will be a big deal is the same reason: because it is so

---

[x] http://stanford.io/1DEUdXl

ABCD Plan: Human Germline Modification
In vitro research is permitted ✔
A. Approval and oversight by SCRO required
B. Bioethics training in advance
C. Clarity and transparency: open access pubs
D. Don't extend to *in vivo* applications ✖

Figure 9.2. The Knoepfler ABCD Plan for managing human genetic modification in the germline. Image source: www.ipscell.com

unsafe. Greely and I disagree on the implications of the safety concerns. He thinks the dangers will discourage people from trying it, while I believe some people will go ahead and do it anyway even though it is unsafe.

ABCD plan

Shortly after the Lanphier and Baltimore pieces, I proposed a plan (Figure 9.2) on my blog to practically manage the situation with human germline genetic modification. I called it the ABCD plan for simplicity.[xi] This plan, which could be implemented in countries around the globe, shares a few key features with some of those already proposed by others, but in some ways it is different or more specific.

Laboratory research on GM human germ cells and early embryos — with appropriate bioethics training and institutional oversight of the scientists involved — can be ethical and of great value if restricted to non-clinical studies. However, it should not be done haphazardly and there must be a compelling reason for doing the gene editing in human embryos versus the less ethically challenging approach of using cultured human cells in a dish limited to the laboratory.

Human embryo editing work done strictly in a laboratory has the potential to provide new, valuable information about gene editing itself. Such research also will likely lead to major discoveries about early human development, fertility, and more. Therefore, such research should not be

[xi] http://www.ipscell.com/abcd-plan/

prohibited, but should only be conducted under certain very limited and strictly controlled conditions in a dish in the labortory.

For example, in my ABCD plan, the *in vitro* studies of GM human germ cells and embryos would require appropriate **approval and oversight**. This mandatory approval in advance is the **A part** of the plan. Most US universities already have committees that oversee stem cell and embryo research (often called "SCRO" for stem cell research and oversight). In the US, these same committees could review applications from researchers wanting to make GM human embryos for non-clinical purposes.

The committees would ideally ask such questions as the following: are the proposed studies ethical and do they need to be done in a human embryo? Realistically, given what I hear, some university committees will simply rubber stamp the work, which is a concern. Also, what do we do in the absence of such committees at institutions around the world that have scientists interested in making GM humans?

Researchers proposing to conduct research on the genetic modification of human germ cells or embryos, to an SCRO, or to any similar committee, must also receive **bioethics training**, which is the **B part of the plan**. This is particularly important because of the complicated bioethical issues that this unique work raises; such training would serve to provide a strong educational component. Bioethical issues to be discussed would include the ethics and research considerations of human embryo modification itself, the specific concerns over outcomes if the work were applied *in vivo*, and other ethical considerations to take into account, such as the sourcing of human oocytes. As to that last issue, *in vitro* CRISPR human genetic modification research could substantially increase the research demand for human eggs. Unfortunately, the vast majority of us scientists receive little to no bioethics training during our careers so we are not necessarily prepared to know with certainty what might be ethical or unethical in this area of research.

The **C part** of the plan is **clarity**. Both the public and scientists would greatly benefit from education and openness in this area. Transparency and outreach in lay terms are essential for building public trust. Research on human germline genetic modification should be published in "open access" publications that make data fully available to society free of charge. This area of research is too important to lock up behind a pay wall of subscription charges. Unfortunately, most research papers that are published, while technically accessible to anyone including the general public, are behind "pay walls." In other words, if you as a non-scientist want to read an article, say on CRISPR gene editing, there is a good chance you will have to pay around $30 to access it or visit your local university library to try to read it. This should not be the case with the human embryo editing papers, as we want the fullest transparency and dialogue.

The **D part** of the plan is **don't extend the work to *in vivo*** applications that involve implanting GM human embryos into women. There should be a prohibition on this step given the major ethical and safety issues involved. The moratorium could potentially be lifted in the future depending on what we learn from the data that emerge from this field in the next few years. Practically speaking, the questions of how such a moratorium would work or be enforced are tough ones, especially if one intends to extend it internationally.

With these **ABCD** guidelines in place, the goal would be that some very specific innovative and valuable research in this area could proceed in a responsible and ethical manner, while minimizing the risk of negative outcomes. Our knowledge of CRISPR-based gene editing will be vastly increased in a few years. Further, in the same timeline, additional next-generation CRISPR approaches will improve accuracy and introduce further refinements to the technology.

This means there is no rush to do clinical gene editing in human embryos; actually, there would be concrete benefits, in the form of enhanced technology and knowledge, to waiting. Let us not forget that for almost all cases of human genetic disease, the safer, alternative approach of using PGD to screen embryos is a powerful, currently available clinical option that should decrease any perceived sense of urgency to start editing human embryos now.

Plans such as my ABCD proposal for managing germline human genetic modification may need to evolve as well. The ongoing dialogue that ramped up in 2015 already shows signs of having had a very positive impact and is likely to continue to do so as research and discussion proceed.

Stanford Law meeting on human genetic modification

I recently participated in a meeting at Stanford Law School on human germline genetic modification hosted by Hank Greely.[xii] The meeting was entitled, "Human Germline Modification: Medicine, Science, Ethics, and Law." In addition to Greely and myself, the panel included the following speakers: Marcy Darnovsky, Executive Director of the Center for Genetics and Society (CGS); Professor Christopher Thomas Scott, Stanford Center for Biomedical Ethics, and Dr. Lynn M. Westphal, Professor of Obstetrics and Gynecology, Stanford University Medical School. You can watch a video of the meeting on YouTube.[xiii]

Part of what made the meeting so valuable was the wide range of views on human modification of the participants and the audience. My sense was that Darnovsky had the strongest concerns about heritable human genetic modification amongst us speakers. CGS, her organization, recently outlined seven reasons on their website to "just say no to human genetic modification"[xiv] and she articulated some of her concerns at the meeting. I share some of her concerns, but do not feel quite as strongly about human embryo editing. Each of us could support *in vitro* research under some conditions.

On the other side, my impression was that Dr. Westphal was perhaps the relatively most open to the possibility of modification being used in the future if deemed safe, particularly if it could help infertile couples have their own genetically-related children. I found it intriguing to hear about Dr. Westphal's experience as a physician on the medical frontline interacting with couples dealing with infertility or genetic problems including mitochondrial disorders. In communicating with Professor Scott afterwards, he indicated that he was struck by the way in which Dr. Westphal indicated that there could well be pressures from patients for therapeutic genetic modification of human embryos.

Greely and Scott were relatively less concerned than Darnovsky about human germline modification overall and more confident that it can be

---

[xii] http://stanford.io/1GcLZoT
[xiii] http://bit.ly/1Be0A3f
[xiv] http://bit.ly/1IFGDSE

regulated appropriately. My impression was that we all could agree that for dealing with most genetic disorders PGD would be far preferable to trying to make a genetic correction in a human embryo.

Greely pointed out three scenarios where PGD likely would not work to correct a problematic genetic situation and hence genetic modification could be needed, which are similar to those I outlined in Chapter 6:

(1) Dominant genetic disease with one parent being homozygous
(2) Both parents have some mutation in the same gene
(3) Certain kinds of mitochondrial disease

Dr. Westphal also added in importantly that for some couples, there are so few embryos that PGD is unlikely to work to find one lacking a mutation. I wondered aloud: how many human embryos might it take to get CRISPR to work for editing too? It might take quite a few.

We had a good discussion of the example of potential genetic modification of a human embryos for the correction of a *BRCA1* breast/ ovarian cancer-associated gene mutation. It was pointed out by several that this mutation is not a "causal" mutation in the sense of always leading to disease, but only confers a predisposition to disease. More broadly there will not always be a clear, bright line between medical use and enhancements with human germline editing.

I thought Professor Scott raised some great points about issues related to scientific publishing and human germline modification. Why did the journal *Protein & Cell*, which published the first human embryo editing paper, only come out with an editorial days later explaining why they decided to publish the controversial paper rather than have an editorial accompanying the actual research article? My sense was that this seemed reactionary and likely was done in response to critical questions raised about the embryo editing paper rather than with forethought about how it would be important to place this publication in an appropriate context right from the beginning.

Some of the questions from the audience at the Stanford meeting included asking about how human modification might change the relationship between parents and children as well as how commercialization could impact the evolution of human editing via CRISPR-Cas9. Another point that came up during the discussion was the concern over potential strong negative impact of gene editing of animals (not humans) either as novelty pets or in the wild.

Overall it was a wonderful, very useful meeting that advanced the discussion of human modification in a positive way. I hope that these types of meetings continue and that there is more discussion of this kind involving diverse stakeholders as well as the public.

I expect that more of these kinds of forums and meetings will take place in the coming months and years. Some may be organized by scientists themselves, while others are likely to be convened by governmental agencies. For instance, the US National Academy of Science is gearing up for meetings on human genetic modification in 2015.[xv]

George Church on human genetic modification

I recently interviewed George Church about the ways that trends in genomics are changing our world and the possibility of heritable human genetic modification. He has a clearly different view than Doudna. Here are excerpts from the interview, originally posted on my blog.[xvi]

Knoepfler: What's your view of 23andMe and other direct-to-consumer genetic testing/genomics companies? I've heard people predict that in the future people will share their genomes on Facebook-like social platforms or even on dating sites. Is there a possibility of the public getting too much genomic/genetic information about themselves or others without enough context? Should kids in school, perhaps as early as primary school, now start to be taught more about their genomes to help prepare the public for these big changes?

Church: Yes. There are now DTC (direct-to-consumer) whole genome sequences for as little as $650,[xvii] and yes, I feel strongly that some

---

[xv] http://bit.ly/1J6Fdl9

[xvi] http://www.ipscell.com/2015/03/georgechurchinterview/

[xvii] http://bit.ly/1KUuYjp

minimum level of genetic literacy should be taught now in every year of school (even if only a few minutes per year). This is a major objective of pgEd.org.[xviii]

Knoepfler: With gene therapy trials, human genetic modification is already ongoing. What about germline human genetic modification? Do you view that as a logical next step?

Church: I don't think that germline is the next goal (nor next logical step), but it might be an acceptable side effect of treating genetic diseases early, safely and effectively. Many gene therapies currently in clinical trials are already aimed at young children to avoid permanent damage. Treating sperm and eggs could reduce the number of abortions (spontaneous and induced) and the number embryos needed in IVF clinics.

Knoepfler: Should we be talking about benefits and risks such as possible future genomics-based eugenics?

Church: Eugenics in USA from 1907 to 1981 involved government sterilization of 65,000 individuals to "improve" the gene pool. The new technology enables parents to make choices about their children just as they might with Ritalin or cleft palate surgery to "improve" behavior or appearance. To prevent such parental decisions, the government would again interfere with reproductive choice, but this time in the apparent opposite direction in terms of improving the gene pool. To give the same name (eugenics) to these two scenarios seems unnecessarily confusing. Should we be talking about benefits and risks? Yes. Frequently and engaging many voices.

Knoepfler: With CRISPR-Cas9 type of technology, the affordability of genomic sequencing, advances in cellular reprogramming, and reproductive cloning, is much more widespread genetic modification of various organisms on our planet inevitable? Desirable? I've read Austen Heinz argue that we should "democratize creation." What do you think of that concept? Will this be extended to human "creation"?

Church: Genetics is not new. Human creativity/creation is not new. These have been "democratized" for centuries, assuming that this means "accessible to many people." Regular folks choose mates for themselves

---

[xviii] http://www.pged.org/

and for other organisms, based on desired traits, and this tends to work (hence the many breeds of animals and plants). CRISPR might make the mutation part a bit easier, but the "selection" part of Darwinian (mutation/selection) is still the key effort.

Knoepfler: I just saw this article[xix] entitled, "Engineering the Perfect Baby" in which Jennifer Doudna was quoted as calling for a temporary moratorium or "pause" on human germ line editing work. Could you please say if you agree or disagree on the call for a pause and explain why in either case?

Church: Note that the word "pause" is not in a quote from Jennifer. I think that Hank, Jennifer and I agree that there should be caution and testing for safety and efficacy for all medical treatments. This inevitably involves "pauses" during pre-clinical and phase 1 clinical trials.

I found this conversation with Church in early 2015 to be a fascinating experience. In advance of interviewing him, I did not imagine the full scope of possibilities presented by the union of transhumanism with genetics and gene editing technology. It is exciting and risky at the same time.

Church, as a very influential scientist both in academia and industry, could lead to wider acceptance of the idea of human genetic modification. At the same time, other scientists pushing back include Doudna as well as the Nobel Laureates David Baltimore and Paul Berg.[xx]

Oxford ethicists: don't worry, just do it!

Over at an online ethics website called "Practical Ethics"[xxi] some prominent Oxford bioethicists have been promoting the idea of going ahead with making GMO sapiens including via a recent blog post there.

The authors of "Editing the germline — a time for reason, not emotion" include Chris Gyngell, Tom Douglas, and Julian Savulescu

---

[xix] http://bit.ly/1aQ7QpI

[xx] http://on.wsj.com/1CWKmbN

[xxi] http://bit.ly/1EYSidy

(Savulescu's pro stance in a 2012 debate on designer babies was discussed in Chapter 6).

They take issue with both the recent *Nature*[xxii] and *Science*[xxiii] perspectives pieces on this topic, and generally want to downplay the potential negative consequences of human germline modification. Some examples of the one-sided verbiage from these bioethicists include the following:

> "Many technologies have unpredictable effects on future generations but this does not mean they are either dangerous or morally unacceptable. Who can predict the effect of information technologies like the internet or smart phones on future generations?... One might object that such technologies don't operate at the genetic level, like CRISPR, and are not passed heritably down to the next generation. But this is a deep mistake — environmental interventions, such as modified social interaction, have epigenetic effects, modify brain development and can be passed on to the next generation... Lasik eye surgery can be used non-therapeutically, but this doesn't justify restrictions on its therapeutic uses... A key fact noticeably absent from both the *Nature* and *Science* commentaries is that many other human activities cause modifications to the human germline."

They argue that creating designer babies with heritable gene edits, is really not so different than, say, having your vision corrected, popping out to smoke a cigarette, changing one's friends, or allowing new generations to use the Internet. This is a false analogy because although these lifestyle choices can in some cases affect the human body, they are intrinsically distinct from heritable DNA changes by genetic design.

Some other bioethicists have made the same sort of argument that common current practices also change our genomes as well, making human genetic modification from their view potentially not so exceptional or worthy of special concern. In his book *The Perfect Baby* [4], bioethicist Glenn McGee also brings forth the idea that we routinely change our children's genomes with certain habits, such as women smoking during

---

[xxii] http://bit.ly/1FfsPOf
[xxiii] http://bit.ly/1IpIope

pregnancy. In theory, a woman who smokes while pregnant could cause her fetus and ultimately the future child that it becomes to inherit one or more genomic mutations due to the mutagenic ingredients of cigarette smoke. Then again, she may not.

The rate of such maternal-induced genetic changes in children is likely to be incredibly low. Even rarer would be mutations in children induced by their mothers that would end up being permanently heritable in that entire future family line. Such mutations would have to occur in the fetus' germ cells so that they would be present in sperm and eggs. Impossible? No, but not likely.

The tobacco-induced mutations, should they occur, are also random in nature and may well occur in regions of the genome that are not of particular importance, such as so-called "gene deserts" where not much happens in the way of genes being active.

In contrast to smoking or other maternal or paternal lifestyle-induced rare mutations, human germline gene editing is a deliberate act, targeted specifically at a certain gene with a focused purpose in mind. This makes gene editing inherently distinct on another level from parental behavior-induced mutations.

The *Practical Ethics* piece promoting human genetic modification unfairly casts opponents as too "emotional" or having other problems. For example, the authors conclude the piece with this rather sharply worded paragraph, which focuses only on the potential upsides and not the downsides of the emerging human modification technology:

> "Gene editing is a revolutionary technology, which potentially offers the next generation an enormous range of benefits. It is important that bad arguments, empty rhetoric and personal interests do not cloud rational thinking and deny the next generation the enormous benefits on offer. It is a time for reason, not emotion."

Thus, to paraphrase them, if one looks at this human germline editing issue logically and without personal bias or excess emotion, one must then necessarily agree with them that it is a good thing to pursue. This unqualified embrace for new technology and "human improvement" provides little in the way of common sense-based reasoning for why designing and making GM humans would be a wise course of action. The debate continues. Later in 2015, an international stem cell and ethics think tank, at which Savulescu is a leading figure, but only one of many members, took a more cautious tone. However, they still were in favour of GM human embryo experiments and there was no mention of a moratorium on clinical use of CRISPR in humans.[xxiv]

The human genome as ever-changing mashup?

There has also been debate over how to view the human genome. Some view it as almost sacred, pure, and not to be tampered with at any cost. On the other hand technophiles in particular think of the human genome as a new frontier to be explored, tweaked, or even profoundly changed for their sense of what is better. To them, the human genome is more like a mutt that is an ever-changing DNA melting pot rather than as something pristine or special. As a result, what is so wrong with changing it more? They argue that we should go ahead and try. Admittedly, I love technology as much as my fellow scientific technophiles. As a genetics researcher I also find it appealing to study the human genome using genetic modification, but only in cells and not in a heritable manner in actual people. It is just too dangerous. Others are more liberal.

For instance, Harvard Professor Steven Pinker has expressed few if any reservations about us diving head first into changing up the human genome. In his view making heritable genetic modifications to the human germline has far more potential for benefits than harms. He generally agrees with the Oxford bioethicists including Savulescu that were discussed in the previous section.

In an August 2015 interview, Pinker indicated to me that he views those who oppose such efforts as being irrational.[xxv] He opposes any moratorium on heritable human modification and in particular singles out bioethicists as obstructing biomedical progress. In his view, they should "get out of the way."

[xxiv] http://bit.ly/1OOEB12
[xxv] http://www.ipscell.com/2015/08/stevenpinker/

When I pointed out to him the fact that prominent scientists such as Professor Jennifer Doudna, Nobel Laureate David Baltimore, and others have strongly and publicly supported a moratorium on heritable human genetic modification, he pulled no punches:

> "The specific harms they warn against, such as inducing cancer, mutations, or birth defects in the unborn child are *already* ruled out by a plethora of existing regulations and norms. *Obviously* we shouldn't mess around with embryos in ways that have a significant probability of producing a sick or deformed child with no compensating benefit. But why do we need a new, across-the-board ban on an entire method to rule out what's already ruled out on the uncontroversial grounds of protecting individuals against foreseeable harm? The authors seem to be acquiescing to the yuck-factor that surrounds the very idea of germline modification, if for no other reason than to draw a firewall around their own research programs, which are restricted to the genetic modification of somatic cells. But scientists should work to dismantle irrational taboos, not indulge them."

This again sounds very reminiscent of the Oxford bioethicists. If you disagree with me, they say, you are irrational or have other problems. In reality, concerns about making GM humans are not simply based on feelings, but rather on logical thought and common sense. Further, to my knowledge there is by no stretch of the imagination any current "ruling out" of the potential harms from making GMO sapiens such as cancer or birth defects. They are actually some of the most likely and dangerous consequences of human genetic modification gone wrong. For example, gene editing has already been proven to cause mutations, both of the desired and undesired kinds.

While the human genome is not some kind of pure, untouchable construct, it nonetheless requires respect as something that is both incredibly powerful and incompletely understood. I share some of the excitement over gene editing, but it is wishful thinking to believe that we can make GM people now or even any time in the next few years without grave risks and today's biomedical regulations are at most only flimsy, outdated protections.

The future of human genetic modification

Where do we go from here?

As we discussed earlier in this book, the required technologies are all either on hand or close to being ready to attempt to make designer babies whether they are clones, GM humans, or both. However, I do not believe science is ready to do it safely and in an ethical manner. Editing the genome of human germ cells or embryos with the intent to try to produce a GM human being would be extremely irresponsible and dangerous at this time and for the fore seeable future. Yet someone will almost certainly try it anyway in coming years.

What happens next?

As discussed in Chapter 4, in the UK another form of trying to make GMO sapiens (three-person IVF) is now legal there and is almost certain to proceed soon. The first GMO sapiens embryo in the UK might be conceived by IVF as early as in a few months. More GM humans also soon may be created in China by another team as well using this IVF kind of technology. At this point there isn't enough information to know if these attempts will succeed or, alternatively, lead to disastrous outcomes.

On the CRISPR front, on December 1–3 2015 a meeting on human genetic modification similar to the one held in 1975 on recombinant DNA technology is planned to take place in the US. At this meeting, some attendees will probably call for a moratorium on the clinical use of genetic modification technology in humans. Then there will be pushback from others against a moratorium on human genetic modification. Some would condemn human genetic modification and the scientists behind it, but those who try to make GM humans would become, in a sense, immortal. That reality is going to be a powerful temptation for some. Others will be driven by curiosity. After all, what could be more cutting edge than making a new type of human being?

Overall, we are left contemplating a rather immediate future where there could be attempts to make GMO sapiens via CRISPR and there will be attempts via three-person IVF technology. Things could get very messy. Remember the attempts to clone sheep and dogs? To make the pioneering animals Dolly and then Snuppy in both cases took many hundreds of tries to get one surviving clone in each case to be born that was at least seemingly healthy.

We should now revisit the type of question asked earlier in the book. If optimistically it takes many (for example, 100) tries to get a single, healthy GMO sapiens child, what happens to the 99 human embryos, fetuses, or even children that did not quite "work out"? Who would be comfortable with that sort of outcome? And what do we do if the GMO sapiens are born apparently healthy, but then are seriously affected by disease later? These truly disturbing possibilities raise the stakes of decisions about this technology.

We need a moratorium now on heritable human genetic modification. I realize that it would be nearly impossible to prevent some people from trying to make GM humans somewhere even with a moratorium in place, but still a moratorium would likely be effective in preventing most attempts at producing GM humans.

Should we go a step further and endorse legislation to legally ban human germline modification in the US? Going beyond a moratorium to new legislative action could pose the risk of doing more harm than good. The reality, for example, that human cloning is legal here in the US today at the federal level (see more in Chapter 3) despite attempts to ban it decades ago is a stark reminder of how difficult it can be to get legislation passed that has any sense of nuance.

Attempts to even temporarily outlaw human genetic modification for clinical purposes in the US could get bogged down as scientists at the same time try to preserve the legality of doing human genetic modification research limited only to cells and test tubes in the laboratory. (I support such research). There will be some lawmakers who will want to ban both experimental and clinical use of human genetic modification technology, while scientists collectively, even if there are some exceptions, will never go for that restrictiveness willingly. They will fight it. I myself would strongly oppose such a law.

If a "clean" federal bill could be passed in the US that only prohibits clinical attempts at human germline modification and does so temporarily, I might be able to support that. However, passage of such a focused bill is very unlikely. And even then many scientists would be against the "clean" bill only banning clinical use because of their hope that human genetic modification technology could potentially become safe and effective some day for clinical use to solve genetic problems that are not addressed by existing technologies, such as PGD. They would argue, probably effectively, that even a temporary legal ban in the US could stifle the path to that exciting future. Note that many countries around the world already ban heritable human genetic modification, but again not the US.

Will heritable human gene editing ever really be a safe and responsible thing to try to accomplish? As of the writing of this book in 2015, I do not foresee scientists having enough information to give it a try ethically and with any reasonable expectation of success within the next several years. In the longer term, it is hypothetically possible, but still a long shot.

Even if at some point gene editing is thought to be perfected so that it always genetically modifies only at the right place and in the right way, there will still be sizable risks. A paradox here is that if clinical use of three-person IVF or CRISPR for the prevention of disease proves successful, which would be wonderful, the downside is that others will take the technology in more questionable or outright unethical directions. Could we limit it only to genetic correction for disease? Almost certainly not. Using such technology for human enhancement could be dangerous to individuals and society. As discussed earlier in this book, Professor Lee Silver (an advocate of making designer babies) imagines a eugenic future with a GMO sapiens upper class that he calls the "GenRich".[xxvi]

Could a new eugenics ideology turbocharged by genetics and the Internet start to permeate society more deeply? Trying to make "better" babies, whatever that means, through genetics would be flirting with disaster. However, it could happen. If one adds in the potential power of disease advocacy, notions of reproductive freedom and parental choice, and libertarianism, the end result could be that human genetic modification becomes mainstream.

To be clear, when the first few new GMO sapiens are born, the world will not suddenly end. Life will go on around the planet. There will be no

---

[xxvi] http://to.pbs.org/1djoeQb

reason for immediate panic because of some imminent move toward a health crisis or possible imminent genetic dystopia. Still, a line will have been crossed and there will be serious risks to society and our species in the longer term.

Does a version of the envisioned genetic future of Professor Silver or of those dystopias of many works of fiction then become possible future realities? Those in favor of technophile, liberal views of human genetic modification argue that such genetic dystopian visions are just cliches. Let's go for it, they say, and not be Luddites. It's easy to be dismissive in that way, but harder to face head on the reality that heritable human genetic modification could eventually spiral out of control with profoundly negative consequences. The latter, sober approach may be less fun to take, but it is more responsible and constructive to positively shaping the future.

Given the lengthy human lifespan and reproductive cycle, one might think that any trend amongst us to increasingly make GMO sapiens would be slow to unfold and therefore provide plenty of opportunity to stop it even after it begins.

However, how can we be sure that we could control it or, if problems do surface, stop it?

The reality is that we just do not know how this trend could develop or if it would be reversible. In only a few decades, for example, we went from IVF baby #1 Louise Brown to over five million IVF babies. That is a good thing and much joy has come out of that technological innovation, but it also serves to highlight how quickly such changes can unfold. If in coming years we witness the birth of GMO sapiens baby #1 made via CRISPR, within a few decades it is conceivable that millions of GM humans could be born, particularly if some genetic modifications become seen as giving children an advantage or are considered to be fashionable. Keep in mind as well that the number of GMO sapiens would multiply as the changes are inherited by each new generation of children.

It could go the other way too with GMO sapiens viewed negatively or even as monsters like some kind of Frankensteins. If they are sickly or have obviously negative traits, they could be ostracized. They could even possibly be prevented from having children through reproductive quarantine or forced sterilization. Some may say that negative eugenics and forced sterilization could never again happen, but how can they be so confident? However it plays out, the GM people produced from such efforts would require respect and legal protection.

We also need to keep in mind the value of human individuality. A great part of the richness of our lives comes from our imperfections, struggles, and differences. There is an irony to the fact that it is these difficulties that both make heritable human genetic modification so seductive (admittedly it resonates with me) and yet also support a powerful argument against it. Striving through genetics toward a likely illusory sense of a perfect existence for our children paradoxically could give their lives less meaning and diversity. And who knows what they would think of being GMO sapiens.

To be clear, a healthier life for the average human being is an appropriate goal for society to strive for, but genetic modification is not the clear path towards that end. We are not ready as a species to take such a road forward and it is possible that we may never be. Still some have said that we should start such efforts now, while others who are relatively more cautious have nonetheless pointed out that in the future it may be considered unethical to not genetically modify our children. I suppose the latter cannot be ruled out, but if that future ever comes to be a reality it is a long, challenging path to get there from where we are today.

What should we do now?

We should educate others and ourselves, catalyze diverse discussions and work to put a hold on attempts at heritable human genetic modification. We should promote at least temporary, national and international moratoriums on clinical use of human germline editing including in the US. While not perfect, these could still substantially limit the extent of attempts at creating new GMO sapiens. In addition this would provide crucial time needed for additional research in the laboratory to broaden our depth of knowledge of genetic modification technology. We also need time to create wise and balanced policies to guide us moving to the future. I have started working on such efforts already through my blog and now via this book.

The confluence of transformative advances in genetic, reproductive, and stem cell technologies are poised to change our world and us with it. Genetic modification technology is part of that. This is a very exciting time to be alive and we should be open to embracing change, but not blindly or in a rush. Armed with information and passion, we can have a major, positive impact on how this biotech revolution unfolds and impacts humanity.

References

1. Lanphier, E, *et al.* (2015) Don't edit the human germ line. *Nature.* **519**: 7544:410–1.
2. Baltimore, BD, *et al.* (2015) A prudent path forward for genomic engineering and germline gene modification. *Science.*
3. Vogel, G (2015) Bioethics. Embryo engineering alarm. *Science.* **347**:6228: 1301.
4. McGee, G (1997) *The Perfect Baby: A Pragmatic Approach to Genetics.* Rowman & Littlefield Publishers.
5. Deonandan, R, S GreenA van Beinum (2012) Ethical concerns for maternal surrogacy and reproductive tourism. *J Med Ethics.* **38**:12:742–5.
6. Araki, M and Ishii, T(2014) International regulatory landscape and integration of corrective genome editing into *in vitro* fertilization. *Reprod Biol Endocrinol.* **12**:108.
7. Suzuki, T, Asami, M and Perry, ACF (2014) Asymmetric parental genome engineering by Cas9 during mouse meiotic exit. *Scientific Rep.* **4**(7621).

Glossary

3-Person IVF. A process intended to prevent mitochondrial disease in humans via either nuclear, spindle, or polar body transfer. Synonymous with "3-parent IVF" or "mitochondrial transfer."

ABCD Plan. The author's proposal for dealing with human genetic modification. The plan includes A for Approval and oversight, B for Bioethics training, C for Clarity and transparency including open access publications of work available at no cost to the public, and D for Don't do clinical work.

Blastomere. A cell from a very early embryo with only eight cells. One of these eight cells is sometimes isolated in a screening or stem cell line production process called "blastomere biopsy" in which the embryo is not destroyed.

Blastocyst. An early stage embryo from which ESC can be isolated that contains a cavity in the middle and which is the most common embryo stage used for IVF.

BRCA1 (Breast Cancer gene 1). A gene that when mutated confers relatively high risk for breast and/or ovarian cancer.

Calgene. A California company that produced the first genetically modified whole food plant product. Eventually bought by Monsanto.

Cas9. A bacterial nuclease enzyme that can cut DNA and becomes specific as part of the CRISPR-Cas9 system.

Cloning. The production of a genetically identical (except for mitochondrial DNA) replicate organism, usually in an asexual manner and most often thought of as occurring in a laboratory setting.

Cystic fibrosis transmembrane conductance regulator (CFTR) gene. The gene whose mutation causes cystic fibrosis.

CRISPR (Clustered Regularly Interspaced Short Palindromic Repeats). Part of the CRISPR-Cas9 system allowing for targeting of specific genomic sequences.

CRISPR-Cas9 system. A bacterial system of defense against pathogens such as viruses that recognizes specific sequences for enzymatic destruction and now adapted for use in producing specific genetic modifications.

De-extinction. The effort to bring back to life an extinct species through reproductive and genetic technologies.

Designer baby. A human baby produced that contains genetic modifications.

Dolly. The first cloned mammal. A sheep cloned from an adult cell of another sheep by Ian Wilmut.

Eggs. In this book usually referring to human oocytes, the female reproductive cells. See also Oocytes.

Embryonic Stem Cells (ESC). Stem cells produced from blastocyst stage embryos that can produce almost any known type of specialized differentiated cell, often by a government.

Eugenics. A movement to improve human beings, implimented either in a positive form (focusing on improving the human genetic state or selective breeding of "better" humans) or a negative form (such as forced sterilization).

FDA (Food and Drug Administration). The government agency in the US that would regulate human genetic modification.

Flvr Savr Tomato. The first genetically modified whole food.

Gene edit. An artificial change in DNA sequence made by scientists. For example, changing a T unit of DNA to an A unit.

Gene Drive. Stimulating biased inheritance of particular genes to alter entire populations.

Gene Drive Chain Reaction. An instance of gene drive going out of control and becoming self-sustaining.

Gene Flow. The transfer of genes between different populations of organisms.

Genetic Determinism. The idea that who we are and how our lives play out to a large extent are controlled by our genetics.

Genetic Modification. Changing of a DNA sequence in a living thing.

Genetic Tourism. People travel from one place to another, usually a destination with weaker regulatory rules, to have a genetic procedure conducted. In the future, this may include production of designer babies.

Genetically Modified (GM). Possessing a genetic modification.

Genetics arms race. Potential cultural or military race between countries to use genetics to create soldiers and perhaps even citizens that are superior. This would be perceived as giving the countries involved an advantage.

Genome. The entirety of an organism's DNA sequence.

GenRich. A GM human upper class imagined by Professor Lee Silver as dominating the future.

GFP (Green Fluorescent Protein). A jellyfish protein that appears green when hit with UV light. Often used as a laboratory tool for genetics studies.

GMO (Genetically Modified Organism). A living thing that has had at least one edit or change to its genomic sequence by another organism.

GMO sapiens. Synonymous with a designer baby.

Guide RNAs. The part of the CRISPR-Cas9 system that guides it to the right place in the genome to make an edit.

Glyphosate. A chemical herbicide to which plant geneticists often engineer resistance in GM plants and seeds. Trade name is Roundup.

Heteroplasmy. A state in which an embryo or organism has mitochondria from at least two different sources rather than the normal single source and associated with negative reproductive outcomes in some cases.

Homo evolutis. The transhumanist terminology for referring to what humanity (*Homo sapiens*) will become after the transformation into something better. At some level synonymous with *GMO sapiens*.

Human Embryology and Fertilisation Authority (HFEA). A UK governmental body tasked with oversight and regulation of research and clinical applications related to human embryos and germ cells.

IPSC (Induced Pluripotent Stem Cells). Stem cells similar in properties to ESC, but made from skin cells or other differentiated cells.

IVF (In Vitro Fertilization). A laboratory process whereby sperm and eggs are mixed in a dish where fertilization occurs outside the body.

Knockout. Animals or cells (most often mice) in which usually a specific single gene has been removed or inactivated.

Mosaicism. A state in which an organism's cells do not all have the same genomic DNA sequence.

Mitochondrial genome. The DNA present in and unique to mitochondria. Also sometimes referred to as Mitochondrial DNA (mtDNA).

Neanderthal. A type of human species now extinct.

Nuclear Transfer. The movement of a somatic cell or oocyte nucleus from one donor cell into a recipient oocyte that has had its own nucleus removed. When cells are produced via this technique they sometimes go by the prefix acronym NT such as in NT-ESC.

Off target effect. An errant CRISPR-Cas9 genome edit.

Oocyte. An egg, such as from a human woman.

Ooplasm. The cytoplasm of an egg.

Ooplasm transfer. The transplant of some egg cytoplasm from one egg to another.

Parthenogenesis. The asexual creation of a cloned organism from only a female parent without male participation.

Polar body. A residual, small, and usually not productive remainder of cell division during oocyte meiosis, but one that contains a full nuclear genome.

Polar body transfer. A method akin to nuclear transfer, but in which the polar body is moved to an egg, which has had its own nucleus removed.

Preimplantation genetic diagnosis (PGD). A procedure in a laboratory in which, one-to-two cells are isolated from early embryos (ideally not destroying the embryos) and analyzed genetically such as for the presence of a disease-causing gene or for sex selection — allowing for creation of a human being with specific traits. This is sometimes also referred to as Preimplantation Genetic Screening (PGS).

Primordial Germ Cells (PGCs). Stem cell-like cells that can differentiate into sperm and eggs.

Raëlians. A UFO cult that believes in human cloning and that has claimed to have cloned people.

Reproductive cloning. The artificial production of a new organism genetically identical to an existing or preexisting one, for example by SCNT.

Reproductive quarantine. Governmental prohibition on reproduction of people and in particular of GMO sapiens perceived to have new negative traits or other heritable problems.

RoundUp. An herbicide to which many GM plants manufactured by Monsanto are resistant.

Savior Sibling. A child whose creation was prompted by parental desire to save an existing child with a disease via some kind of transplant from the new offspring to the older sibling.

SCNT (Somatic Cell Nuclear Transfer). The movement of a non-germ cell or its nucleus into an oocyte that has had its own nucleus removed.

Sex selection. The use of PGD for a parent or parents to produce a human offspring that is specifically male or female.

Spindle Transfer. The same as nuclear transfer except rather than the entire nucleus, only the chromatin spindle is transplanted.

TALENS. A tool for gene editing, similar to CRISPR-Cas9 but more complicated and expensive.

Therapeutic cloning. The creation of ESC lines from another organism, usually an adult.

Three-person IVF. The technique of nuclear transfer combined with IVF to try to produce children (who would be genetically modified) without mitochondrial diseases. Also known as three-parent IVF and mitochondrial transfer or donation.

Transhumanism. A movement with the purpose of using technology to catalyze the transcendence of humanity to become a new, better species.

Index

GMO sapiens : the life-changing science
of designer babies / Paul Knoepfler,
University of California, Davis.